Screw Compressors

A. Kovacevic N. Stosic I. Smith

Screw Compressors

Three Dimensional Computational Fluid Dynamics
and Solid Fluid Interaction

With 83 Figures

Ahmed Kovacevic
Nikola Stosic
Ian Smith

School of Engineering and Mathematical Sciences
City University
Northampton Square
London
EC1V 0HB
United Kingdom

Library of Congress Control Number: 2006930403

ISBN 3-540-36302-5 Springer-Verlag Berlin Heidelberg New York

This work is subject to copyright. All rights are reserved, whether the whole or part of the material is concerned, specifically the rights of translation, reprinting, reuse of illustrations, recitation, broadcasting, reproduction on microfilm or in any other way, and storage in data banks. Duplication of this of this publication or parts thereof is permitted only under the provisions of the German Copyright Law of September 9, 1965, in its current version, and permission for use must always be obtained from Springer-Verlag. Violations are liable for prosecution under the German Copyright Law.

Springer-Verlag is a part of Springer Science+Business Media

© Springer-Verlag Berlin Heidelberg 2007
springer.com

The use of general descriptive names, trademarks, etc. in this publication does not imply, even in the absence of a specific statement, that such names are exempt from the relevant protective laws and regulations and therefore free for general use.

Cover-Design: Estudio Calamar, Viladasens
Typesetting by the authors and SPi

Printed on acid-free paper   SPIN: 11741589   62/3100/SPi   5 4 3 2 1 0

Preface

Screw compressors are rotary positive displacement machines, which are compact, have few moving parts and which operate at high efficiency over a wide range of speeds and pressure differences. Consequently a substantial proportion of all industrial compressors now produced are of this type.

There are few published books on the principles of their operation and how best to design them, especially in English. The authors made a first step to fill this void with an earlier work on mathematical modelling and performance calculation of these machines. This described analytical procedures which are generally adequate for most applications, especially when dealing with oil flooded machines, in which temperature changes are relatively small and thus the effect of changes of shape of the key components on performance may be neglected. This assumption permitted the use of analytical procedures based on flow through passages with dimensions that are invariant with temperature and pressure.

As manufacturing accuracy increases, clearances can be reduced and compressors thereby made smaller and more efficient. To obtain full advantage of this at the design stage, more accurate procedures are required to determine the internal fluid flow patterns, the pressure and temperature distribution and their effects on the working process. This is especially true for oil free machines, in which temperature changes are much larger and thus make thermal distortion effects more significant.

The present volume was prepared, as a sequel to the authors' earlier work and describes the most up to date results of methods, which are still being developed to meet this need. These are based on the simulation of three-dimensional fluid flow within a screw machine.

Such an approach requires the simultaneous solution of the governing equations of momentum, energy, mass, concentration and space conservation. In order to be solved, the equations are accompanied by constitutive relations for a Newtonian fluid in the form of Stokes, Fourier's and Fick's laws. Although this model is generally applicable, at least three features have been derived, without which the solution of the flow pattern within a screw machine would not be possible. Firstly, there is the application of the Euler-Lagrange method for the solution of multiphase flow of both the injected fluid and the liquid phase of the main fluid. Its equations define mass, energy and concentration source terms for the main equations. The introduction of a 'boundary domain' in which the pressure is kept constant by injection or by subtraction of mass, is the next innovation in and allows simulation of the real operational mode of a screw compressor. Finally, a rather simple method is used here for calculation of the properties of real fluids. This permits a fast but still reasonably accurate procedure even for the calculation of processes like evaporation or condensation within the working chamber.

A computer program has been developed for Windows or UNIX oriented machines, based on the methodology, thus described. It automatically generates files which contain node, cell and region specifications as well as user subroutines for a general CFD solver of the finite volume type. The method and program have been tested on a commercial CFD solver and good simulations of positive displacement screw machines were obtained.

Four examples outline the scope of the applied mathematical model for three dimensional calculation of fluid flow and stresses in the solid parts of the screw machine. In the first example, the results for two oil-free air screw compressors with different rotor profiles are compared with each other and with results obtained from one-dimensional calculation. Advantages are found in the use of the three-dimensional simulation model, in which the suction and discharge dynamic losses are accounted for in the results.

The second example verifies the results of three-dimensional calculations with measurements obtained in an experimental test rig. The influence of turbulence upon the process in a screw compressor is also investigated. It is concluded that, despite the excessive dissipation of kinetic energy of turbulence, the overall parameters of a positive displacement machine do not significantly change if it is calculated either as turbulent or laminar flow. Investigation of the grid size influence on the accuracy of 3-D calculations is performed and it shows that coarser numerical meshes cannot capture all of the flow variations within the compressor accurately. However, the integral parameters are in all cases reasonably accurate.

In the third example the procedures for real fluid properties enabled fast calculation of ammonia compressor parameters. The importance of the oil injection port position is outlined here through the oil distribution obtained by the three dimensional calculations. Such figures of oil distribution within the screw compressor have never been found in the open literature. This achievement is therefore a step forward in understanding screw compressor internal flows.

The fourth application covers the simultaneous calculation of the solid fluid interaction. The influence of the rotor deformation on the integral screw compressor parameters caused by the change in clearance is presented together with how rotor clearances are reduced due to the enlargement of the rotors caused by temperature dilatations. This results in an increase in both, the compressor flow and power input. The influence of pressure causes the rotors to bend. For a moderate compressor pressure, the clearances gap is enlarged only slightly and hence has only a negligible influence on the delivery and power consumption. In the case of high working pressures, the rotors deform more and the decrease in the delivery and rise in specific power becomes more pronounced.

Ahmed Kovacevic
Nikola Stosic
Ian Smith

London,
February 2006

Notation

A	- area of the cell surface
$\mathbf{a}_1, \mathbf{b}_1$	- radius vectors of boundary points
A_1, A_2	- constants in the saturation temperature equation
b_i	- constant
B_1, B_2	- constants in the compressibility factor equation
c_i	- concentration of species
c_1-c_4	- tension spline coefficients
C_1, C_2	- coefficients of the orthogonalisation procedure
C_{drag}	- drag coefficient
C_p	- specific heat capacity at constant pressure
C, σ	- constants in k-ε model of turbulence
d_i	- distance in transformed coordinate system
D_i	- mass diffusivity of the dispersed phase
d_o	- Sauter mean droplet diameter
D_1-D_4	- constants in the vapour specific heat equation
$\dot{\mathbf{D}}$	- rate of strain tensor
e_1, e_2	- cell edges maximal values in coordinate directions
$f(s)$	- adaptation variable
$\mathbf{f}_b$	- resultant body force
f_e	- expansion factor
f_k	- weight function
$F^i(s)$	- integrated adaptation variable
h_1-h_8	- blending functions of Hermite interpolation
h	- enthalpy
h_L	- enthalpy of evaporation
k	- turbulent kinetic energy
K_1-K_4	- coefficients for Hermite interpolation
k_P	- point counter
K	- number of points
m	- mass
m_o	- mass of species
m_i	- mass in the numerical cell
Δm_L	- mass of evaporated or condensed fluid
n	- rotor speed
Nu	- Nuselt number
p	- pressure
P	- production of turbulence energy

VIII Notation

Pr	- Prandtl number
$\mathbf{q}_{ci}$	- diffusion flux of species
$\mathbf{q}_h$	- heat flux
$\mathbf{q}_k$	- diffusion flux in kinetic energy equation
$\mathbf{q}_\varepsilon$	- diffusion flux in dissipation equation
$\mathbf{q}_{\phi S}$	- flux source in the generic transport equation
$\mathbf{q}_{\phi V}$	- volume source in the generic transport equation
$\dot{Q}_{con}$	- convective heat flux
$\dot{Q}_{mass}$	- heat flux due to phase change
$\mathbf{r}$	- radius vector
Re	- Reynolds number
R_i	- grid point ratio
S	- cell surface
s	- transformed coordinate
$\mathbf{s}$	- area vector
S_{ci}	- source term of species
S_h	- heat source term
$\mathbf{S}$	- viscous part of stress tensor
t	- time
T	- temperature
$\mathbf{T}$	- stress tensor
$\mathbf{u}$	- displacement vector
$\mathbf{v}$	- fluid velocity
$\mathbf{v}_{ci}$	- velocity of the dispersed phase
v_i	- Cartesian component of velocity vector
$\mathbf{v}_s$	- surface velocity
V	- cell volume
V_{CM}	- volume of the control mass
V_{CV}	- control volume
w	- weight factor
W	- weight function
X_ξ	- grid spacing
x, y, z	- physical coordinates
X, Y, Z	– points on physical boundaries
x_p, y_p	- coordinates of the calculation point
y_p', y_p''	- first and second derivatives in vicinity of the point P
y^+	- dimensionless distance from the wall
z	- compressibility factor

Greek symbols

α_t	- linear thermal expansion coefficient
α, β	- tension spline coefficients
α_1, β_1	- blending functions
δ	- Kroneker delta function
δt	- time step
ε	- dissipation of turbulent kinetic energy
ϕ	- transported property in generic transport equation
φ	- interlobe angle of male rotor
Γ_ϕ	- Diffusive term in generic transport equation
η, λ	- Lame coefficients
κ	- thermal conductivity
μ	- viscosity
μ_t	- turbulent viscosity
π	- the ratio of a circle's circumference to its diameter
ω	- angular velocity
ρ	- density
σ	- tension spline parameter
σ_o	- oil surface tension
σ_{cv}	- normalised cell volume
ξ, η	- computational coordinates
$\hat{\xi}_o, \hat{\xi}_1$	- one-dimensional stretching functions
$\hat{\xi}$	- multi-dimensional stretching function

Subscripts

1	- male rotor
2	- female rotor
add	- injected / subtracted fluid
$const$	- constant prescribed value
D	- discharge bearings
i	- dispersed phase
L	- evaporated/condensed fluid
l	- liquid
M	- mixture
max	- maximum
min	- minimum
o	- oil
ref	- reference value
s	- grid values
S	- surface

X   Notation

sat   - saturation
t   - turbulence
v   - vapour
V   - volume

Superscripts

'   - fluctuating components for time averaging
"   - fluctuating component for density-weighted averaging
k   - number of time steps

Contents

1 Introduction 1
    1.1 Screw Machines 1
    1.2 Calculation of Screw Machine Processes 3
    1.3 Fluid Flow Calculation 3

2 Computational Fluid Dynamics in Screw Machines 7
    2.1 Introduction 7
    2.2 Continuum Model applied to Processes in Screw Machines 8
        2.2.1 Governing Equations 9
        2.2.2 Constitutive Relations 12
        2.2.3 Multiphase Flow 14
        2.2.4 Equation of State of Real Fluids 19
        2.2.5 Turbulent Flow 22
        2.2.6 Pressure Calculation 23
        2.2.7 Boundary Conditions 23
    2.3 Finite Volume Discretisation 27
        2.3.1 Introduction 27
        2.3.2 Space Discretisation 29
        2.3.3 Time Discretisation 29
        2.3.4 Discretisation of Equations 30
    2.4 Solution of a Coupled System of Nonlinear Equations 32
    2.5 Calculation of Screw Compressor Integral Parameters 33

3 Grid generation of Screw Machine Geometry 39
    3.1 Introduction 39
        3.1.1 Types of Grid Systems 42
        3.1.2 Properties of a Computational Grid 44
        3.1.3 Grid Topology 47
    3.2 Decomposition of a Screw Machine Working Domain 48
    3.3 Generation and Adaptation of Domain Boundaries 52
        3.3.1 Adaptation Function 53
        3.3.2 Adaptation Variables 55
        3.3.3 Adaptation Based on Two Variables 55
        3.3.4 Mapping the Outer Boundary 58
    3.4 Algebraic Grid Generation for Complex Boundaries 62
        3.4.1 Standard Transfinite Interpolation 63

      3.4.2 Ortho transfinite interpolation 65
      3.4.3 Simple Unidirectional Interpolation 73
      3.4.4 Grid Orthogonalisation 75
      3.4.5 Grid Smoothing 78
      3.4.6 Moving Grid 79
   3.5 Computer Program 81

4 Applications 83
   4.1 Introduction 83
   4.2 Flow in a Dry Screw Compressor 84
      4.2.1 Grid Generation for a Dry Screw Compressor 86
      4.2.2 Mathematical Model for a Dry Screw Compressor 87
      4.2.3 Comparison of the Two Different Rotor Profiles 87
   4.3 Flow in an Oil Injected Screw Compressor 94
      4.3.1 Grid Generation for an Oil-Flooded Compressor 97
      4.3.2 Mathematical Model for an Oil-Flooded Compressor 97
      4.3.3 Comparison of the Numerical and Experimental results for an Oil-Flooded Compressor 98
      4.3.4 Influence of Turbulence on Screw Compressor Flow 106
      4.3.5 The Influence of the Mesh Size on Calculation Accuracy 114
   4.4 A Refrigeration Compressor 118
      4.4.1 Grid Generation for a Refrigeration Compressor 118
      4.4.2 Mathematical Model of a Refrigeration Compressor 119
      4.4.3 Three Dimensional Calculations for a Refrigeration Compressor 119
   4.5 Fluid-Solid Interaction 123
      4.5.1 Grid Generation for Fluid-Solid Interaction 123
      4.5.2 Numerical Solution of the Fluid-Solid Interaction 124
      4.5.3 Presentation and Discussion of the Results of Fluid-Solid Interaction 125

5 Conclusions 131

A Models of Turbulent Flow 133

B Wall Boundaries 139

C Finite Volume Discretisation 143

References 155

1
Introduction

1.1 Screw Machines

The operating principle of screw machines, as expanders or compressors, has been known for over 120 years. Despite this, serious efforts to produce them were not made until low cost manufacturing methods became available for accurate machining of the rotor profiles. Since then, great improvements have been made in performance prediction, rotor profile design and manufacturing techniques. Screw compressors are now highly efficient, compact, simple and reliable. Consequently, they have largely replaced reciprocating machines for the majority of industrial applications and in many refrigeration systems.

Screw compressors and expanders are positive displacement rotary machines. They consist essentially of a pair of meshing helical lobed rotors, which rotate within a fixed casing that totally encloses them, as shown in Figure 1-1.

Figure 1-1 Screw Compressor Components

Although screw machines can function as either expanders or compressors, their overwhelmingly common use is as compressors, of which there are two main types. These are oil flooded, commonly known as oil injected, and oil free compressors. An example of each, with similar rotor sizes, is shown in Figure 1-2.

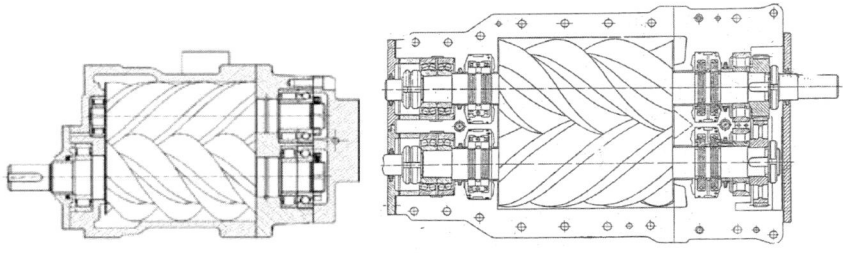

a) Oil Injected Compressor   b) Oil Free Compressor

Figure 1-2 Types of Screw Compressor

In oil injected compressors, a relatively large mass, though a very small volume, of oil is admitted with the gas to be compressed. The oil acts as a lubricant between the contacting rotors, a sealant of any clearances between the rotors and between the rotors and the casing and as a coolant of the gas during the compression process. This cooling effect improves the compression efficiency and permits pressure ratios of up to approximately 15:1 in a single stage, without an excessive temperature rise, by maintaining an oil:gas mass ratio of 4:1 or even more. The effects of thermal expansion are then relatively small and now that screw compressor components can be manufactured with tolerances of the order of $\pm 5\mu m$, internal clearances can be as little as 30-60μm.

In oil free compressors, only gas is admitted into the working chamber. External timing gear is therefore needed, in order to prevent rotor contact, and internal shaft seals have to be located between the bearings at each end of the rotor shaft and the main body of the rotor. The shaft seals are needed to prevent oil, which is supplied to the bearings through an external lubrication system, from entering the working chamber and thereby contaminating the gas being compressed. Because there is no injected oil to cool the gas in this type of machine, the temperature rise of the compressed gas is much higher than in oil flooded compressors and pressure ratios are therefore limited to approximately 3:1, depending on the type of gas being compressed. Above this value the temperature rise associated with compression creates problems related to rotor and casing distortions. Clearances therefore have to be much larger in order to avoid contact between the rotors or between the rotors and their casing caused by thermal distortion. It is believed that the adiabatic efficiency of oil free compressors could be increased by as much as 10%, if minium safe clearances could be predicted accurately.

1.2 Calculation of Screw Machine Processes

Up till approximately 1980, screw compressors were designed assuming an ideal gas in a leak proof working chamber going through a compression process which could reasonably be approximated in terms of pressure-volume changes by the choice of a suitable value of exponent "n" in the relationship pV^n = Constant.

To improve on this procedure it was first necessary to obtain an algorithm which could be used to estimate the trapped volume between the rotors and the casing and the areas of all leakage passages, at any rotational angle. The latter are formed by clearances between the rotors and between the rotors and the casing. In addition, the area of the inlet or exit passage exposed to bulk flow of fluid into or out of the working chamber had to be obtained where applicable.

The assumption of dimensionless non-steady bulk fluid flow and steady one dimensional leakage flow through the working chamber, together with suitable flow coefficients through the passages and an equation of state for the working fluid, made it possible to develop a set of non-linear differential equations which describe the instantaneous flow of fluid, work and heat transfer through the system. These could be solved numerically to estimate pressure-volume changes through the suction, compression and delivery stages and hence determine the net torque, power input, fluid delivery and isentropic and volumetric efficiencies of a compressor. More details of this are given in the authors' earlier volume *Stosic, Smith and Kovacevic,* (2005).

1.3 Fluid Flow Calculation

In recent years there has been a steady growth in the use of Computational Fluid Dynamics (CFD) as a means of calculating 3-D external and internal flow fields. It is widely used today for estimating flow in rotating machinery and specialised codes have been developed for this to allow faster calculations.

Many books on fluid dynamics such as *Bird et al* (1962), *Fox and McDonald* (1982) and *White* (1986) contain a detailed derivation of general conservation laws. Three main groups of methods have been developed through the years as described by *Ferziger and Perić* (1995). These are the finite difference, finite element and finite volume methods. It is believed that the finite difference was first described by *Euler* in the 18th century but, more recently, *Smith* (1985) gave a comprehensive account of all its aspects.

The finite element method was initially developed for structural analysis, but later has also been used for the study of fluid flow. It has been described extensively by many authors, such as *Oden* (1972), *Fletcher* (1991) and *Zienkiewicz and Taylor* (1991).

A summary of the finite volume method is given by *Versteeg and Malalasekera* (1995). Since the finite volume method has already been used to solve problems involving unsteady flow with moving boundaries and strong pressure-velocity-density coupling, it is of particular interest for this book. The 'space conservation law' was introduced by *Trulio and Trigger* (1961) and used in conjunction with a finite difference method. The importance of the space conservation law was discussed by *Demirdžić and Perić* (1988) and introduced to the finite volume method for prediction of fluid flow in complex domains with moving boundaries by the same authors (1990) and also by *Demirdžić, Issa and Lilek* (1990). They followed the attempts of many other authors to apply it to solve some special cases. Typically, *Gosman and Watkins* (1977), *Gosman and Johns* (1978) and *Gosman* (1984) reported the calculations of flows in a cylinder with moving boundaries. *Stošić* (1982) applied the method to internal unsteady flows of a compressible fluid. *Thomas and Lombard* (1979) presented solutions of steady and unsteady supersonic flows while *Gosman* (1984) and *Durst et al* (1985) reported that simple transformation of the conservation equations enables easy discretisation when only one of the domain boundaries moves in one direction. *Perić* (1985) introduced a finite volume methodology for prediction of three-dimensional flows in complex ducts where, among others, he gave an evaluation of various pressure-derivation algorithms for orthogonal and non-orthogonal grids. Additional analysis on pressure-velocity coupling is given by *Perić* (1990) and later discussed and improved by *Demirdžić at al* (1992) and *Demirdžić and Muzaferija* (1995) where they applied the method simultaneously to fluid flow and solid body stress analysis. Turbulence modeling is discussed by many authors, among whom *Hanjalić* (1970) gave an essential introduction to its wider use. *Bradshaw* (1994) and *Hanjalić* (1994) gave good summaries on the subject. *Lumley* (1999) outlined important subjects on turbulence in internal flows of positive displacement machines.

Despite a large number of publications on CFD, little has been written on its use for the analysis of flow through screw machines. This is mainly due to the complexity of both, the machine configuration and the flow paths through them. Some existing commercial CFD codes have facilities that can cope with the complex geometry of screw machines. Unfortunately, these codes need to be improved in order to give useful results. In addition, a pre-processor needs to be developed to generate a numerical mesh that describes their shape with sufficient accuracy and allows for the complex stretching and sliding motion associated with the flow.

With the advance of computing, it is now possible to predict internal flow in screw compressors by 3-D methods so that the internal pressure and temperature distribution can be estimated throughout the machine. Further, this can be used as a basis for determining the distribution of injected oil in oil flooded machines, as a means to estimate thermal distortion within an oil free compressor and to design inlet and exit port passages with minimum flow losses.

Such a procedure makes it possible to reduce the size of screw compressors by bringing internal leakage to a minimum. This would improve the adiabatic efficiency of such machines by virtue of the reduced internal losses and greatly reduce the cost of developing new products by cutting the time and cost of experimental testing and development.

Despite a significant increase in the number of papers published recently in the area of computational fluid dynamics, only a few deal with the application of computational fluid dynamics to screw compressors. All of them are recent papers by *Kovacevic, Stosic and Smith* published between 1999 and 2005. These papers introduced 3-D numerical analysis to the screw compressor world. Their latter papers are related to both, grid generation in screw compressors and 3D numerical performance estimation, *Kovacevic et al* (2003) and (2005). These include fluid solid interaction in screw machines, *Kovacevic et al* (2004).

2

Computational Fluid Dynamics in Screw Machines

2.1 Introduction

As computer technology and its associated computational methods advance, the use of 3-D Computational Fluid Dynamics (CFD) to design and analyse positive displacement machinery working processes is gradually becoming more practicable. In general, the CFD modelling process can be split into four phases.

The first phase is concerned with defining the problem that has to be solved. Both the ease of solution and implementation of results into the design process are heavily dependent on this critical starting step. Two different approaches are available for screw machines. The first is to select one interlobe on the main rotor and the corresponding interlobe on the gate rotor in order to make a computational domain. This is probably the easiest to implement but takes no account of important phenomena such as interlobe leakage, blow-hole losses, oil injection and oil distribution. Another approach assumes that the whole domain of a screw machine is analysed. This includes the suction chamber and its port, the compression or expansion chamber with its moving rotor boundaries and the discharge system of the machine. By this means, the leakage paths and any additional inlet or outlet ports are included in the domain to be analysed. Realism in representing the machine working process gives a large advantage to this approach. The design procedure and the CFD numerical analysis can then be easily connected and interchanged and the calculation of the operational parameters of such machines is thereby facilitated. Unfortunately, such a complex geometry cannot be represented by a small number of computational points.

In the second phase, a mathematical model that is capable of describing the problem has to be selected. There are again two types of situation. The first is where an adequate mathematical description exists and can be used, e.g. heat conduction, elastic stress analysis and laminar fluid flow. The second is where such a description either does not exist or is impracticable to use, e.g. non-linear stress analysis and turbulent fluid flow. In the case of positive displacement machines, it is unlikely that any analytical solutions exist. This is because highly compressible flow appears inside both domains with turbulent flow regimes and domains with low Reynolds numbers. There is additional non-linearity introduced by two-phase flow, particle flow, moving and stretching domains and sliding boundaries. Due to all these, the mathematical model implemented here needs to cope with a variety of different requirements. It is based on the general laws of mass, momentum, energy and space conservation. The resulting system of governing equations is not closed because it contains more unknowns than resulting equations. It is closed by constitutive relations, which give information about the response of a particular continuum material to external influences. The whole concept of mathematical modelling is based on a phenomenological approach which employs the principle of a continuum as the physical background. It can be applied only when an elementary part of material or the smallest characteristic length of the flow, which has to be analysed, is much bigger then the mean free molecular path. Fortunately, this condition is fulfilled for the majority of fluids and practically for all solid structures.

The mathematical description of problems in continuum mechanics is very rarely amenable in a closed-form of analytical solution and an iterative numerical procedure is thus the only alternative that can be applied to solve models in positive displacement machines. Numerical methods transform the differential equations of the mathematical model into a system of algebraic equations. The third phase is therefore to select the discretisation method. To do that a number of approximations are made: the continuum is replaced by a set of computational points with finite distances between them in space and time, while the continuous functions which represent the exact solution of the mathematical model are approximated by polynomials, typically of a second order. Because of the complexity of positive displacement machines, the standard approach to spatial discretisation is not applicable and a special grid generation method has to be developed and applied to them. The equations are discretised by the finite volume method, which appears to have a more conservative form of governing laws then any other numerical method. The result of discretisation is a system of algebraic equations the size of which depends on the number of numerical cells.

The resulting set of algebraic equations is then solved by approximate iterative methods. Iterations are necessary due to the non-linearity of the mathematical model. Even for linear problems, an iterative solution method is usually more efficient than a direct one. In addition, iterative solution methods are less sensitive to round-off errors due to the finite accuracy of the computer arithmetic.

2.2 Continuum Model applied to Processes in Screw Machines

A mathematical model of the transport processes, which exist within both twin screw and other types of positive displacement machine, is presented here. It includes the mass, momentum and energy conservation equations in integral form, a space conservation law, which has to be satisfied for problems with a moving mesh, constitutive relations required for the problem closure, a model of dispersed flow, models of turbulence in fluid flow and boundary conditions.

All the equations are presented in a symbolic coordinate-free notation which directly conveys the physical meaning of particular terms without unnecessary reference to any coordinate system. However, numerical solution of these equations requires a coordinate system and vectors and tensors have to be specified in terms of their components.

2.2.1 Governing Equations

Fluid contained within a screw compressor can be gas, vapour or a wet mixture of liquid and vapour. In some cases, it can be pure liquid. Its density varies with both pressure and temperature. The compressor flow is governed by equations based on the general laws of continuity, momentum and energy conservation. The most general approach is to write these equations in integral form and apply them to an arbitrary part of the fluid or solid continuum of volume V, which is bounded by a moving surface S, as shown in Figure 2-1.

Reynolds' transport theorem can be expressed as:

$$\frac{d}{dt}\int_{V_{CM}} \rho\phi dV = \frac{d}{dt}\int_{V_{CV}} \rho\phi dV + \int_{S} \rho\phi(\mathbf{v}-\mathbf{v}_s)d\mathbf{s} \tag{2.1}$$

where, V_{CM} is the volume of the control mass, V_{CV} is the control volume enclosed by the surface S. Vector d**s** stands for the outward pointing surface vector, defined by its unit vector **n** and surface area dS as d**s**=**n** dS. In equation (2.1), ϕ represents any intensive property based on mass, momentum, energy, concentration or other parameter.

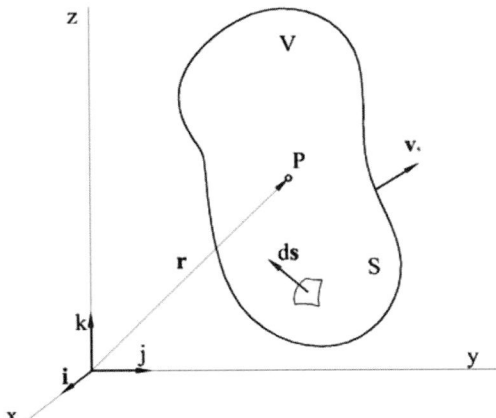

Figure 2-1 Control volume of part of the continuum

Vector $\mathbf{v}_s$ is the velocity with which the surface of the control volume moves. If the control volume is fixed, then its surface velocity $\mathbf{v}_s=0$. Equation (2.1) then makes the rate of change of the amount of property in the control mass equal to the sum of the rate of change of that property within the control volume and its net flux through the control volume boundary due to relative fluid motion. If the control volume moves with the same velocity as the boundary of the control mass, then the boundary velocity is equal to the velocity of the control mass, $\mathbf{v}=\mathbf{v}_s$. For convenience, the control volume is denoted as V and its surface as S.

If the variable ϕ in equation (2.1) has the value of 1, then the equation represents that of continuity:

$$\frac{d}{dt}\int_V \rho dV + \int_S \rho(\mathbf{v}-\mathbf{v}_s)\cdot d\mathbf{s} = 0, \qquad (2.2)$$

If the conserved property is velocity, i.e. $\phi=\mathbf{v}$, then equation (2.1) becomes that of the conservation of momentum:

$$\frac{d}{dt}\int_V \rho \mathbf{v} dV + \int_S \rho \mathbf{v}(\mathbf{v}-\mathbf{v}_s)\cdot d\mathbf{s} = \sum f \qquad (2.3)$$

where the right hand side of the equation represents the sum of surface and body forces which act on the matter in the control volume. Since the body forces acting on the whole matter trapped in the control volume are independent of the shape of the boundary surface, they represent a vector field and can be integrated over the control volume. However, surface forces such as pressure forces, normal and shear stress forces or surface tension forces, depend on the surface on which they act, and they represent momentum fluxes across the surface. More details of this can be found in *Ferziger and Peric* (1995). In order to close the system of equations, these fluxes must be written in terms of properties whose conservation is governed by the equation in question. In equation (2.3) the conserved property is the velocity $\mathbf{v}$. For Newtonian fluids, a constitutive relation between stress $\mathbf{T}$ and strain $\mathbf{D}$ is Stokes' law. Hookes' law gives a constitutive relation for thermo-elastic solids. The momentum equation (2.3) then becomes:

$$\frac{d}{dt}\int_V \rho \mathbf{v} dV + \int_S \rho \mathbf{v}(\mathbf{v}-\mathbf{v}_s)\cdot d\mathbf{s} = \int_S \mathbf{T}\cdot d\mathbf{s} + \int_V \mathbf{f}_b dV \qquad (2.4)$$

where $\mathbf{T}$ is the stress tensor and $\mathbf{f}_b$ is the resultant body force.

If the conserved property ϕ in equation (2.1) is scalar, then the equation can be written in the following form:

$$\frac{d}{dt}\int_V \rho \phi dV + \int_S \rho \phi(\mathbf{v}-\mathbf{v}_s)\cdot d\mathbf{s} = \sum f_\phi, \qquad (2.5)$$

where the term on the right hand side is the sum of all the modes of transport of the property ϕ, other than convection, which is already on the left side of this equation, and any sources or sinks of that property. This sum generally consists of

2.2 Continuum Model applied to Processes in Screw Machines

two terms; the diffusive transport and the sink or source of the conserved property. The diffusive transport is:

$$f_\phi^d = \int_S \Gamma \, \text{grad} \phi \cdot d\mathbf{s} = \int_S \mathbf{q}_\phi \cdot d\mathbf{s} \, . \tag{2.6}$$

Γ is the diffusivity of ϕ. Equation (2.5) in that case becomes a general conservation equation:

$$\frac{d}{dt} \int_V \rho \phi dV + \int_S \rho \phi (\mathbf{v} - \mathbf{v}_s) \cdot d\mathbf{s} = \int_S \mathbf{q}_\phi \cdot d\mathbf{s} + \int_V S_\phi dV \tag{2.7}$$

where S_ϕ is the source or sink of property ϕ per unit mass. Equation (2.7) appears to be a generic equation valid for all intensive properties of matter.

From equation (2.7) one can get the energy equation, in the form of enthalpy, directly as:

$$\frac{d}{dt} \int_V \rho h dV + \int_S \rho h (\mathbf{v} - \mathbf{v}_s) \cdot d\mathbf{s} = \int_S \mathbf{q}_h \cdot d\mathbf{s} + \int_V s_h dV +$$
$$\int_V (\mathbf{v} \cdot \text{grad} \, p + \mathbf{S} : \text{grad} \, \mathbf{v}) dV - \int_S p\mathbf{v}_s \cdot d\mathbf{s} + \frac{d}{dt} \int_V p dV \tag{2.8}$$

S is the viscous part of the stress tensor:

$$\mathbf{S} = \mathbf{T} + p\mathbf{I} \tag{2.9}$$

I is a unit tensor.

If applied to the concentration scalar $c_i = {m_i}/{m}$, where m_i denotes the mass of the dispersed fluid in the working fluid and m defines the overall mass, equation (2.7) becomes:

$$\frac{d}{dt} \int_V \rho c_i dV + \int_S \rho c_i (\mathbf{v} - \mathbf{v}_s) \cdot d\mathbf{s} = \int_S \mathbf{q}_{c_i} \cdot d\mathbf{s} + \int_V S_{c_i} dV \, , \tag{2.10}$$

where $\mathbf{q}_{ci}$ is the diffusion flux and S_{ci} is the source or sink of the dispersed phase.

If the conserved property in equation (2.7) is defined as $\phi = 1/\rho$ then this equation becomes the space conservation law which must be satisfied in all cases even if the domain boundaries move:

$$\frac{d}{dt} \int_V dV - \int_S \mathbf{v}_s \cdot d\mathbf{s} = 0 \, . \tag{2.11}$$

This equation links the rate of change of volume V and surface the velocity $\mathbf{v}_s$.

Equations (2.2), (2.4), (2.8), (2.10) and (2.11) constitute a mathematical model which is valid for the majority of fluids and solids in engineering practice. For the numerical modelling of a screw machine, the first three of these equations should be solved for the working gas or vapour, which is a background fluid. Equation (2.10) is solved for the disperse phase, which is either oil or other fluid injected into the working chamber and dispersed into the background fluid, while the equation of space must be satisfied for any case, because the compression or expansion in a positive displacement machine is caused entirely by the movement of the boundary. In two-phase flow, the liquid phase of the working fluid can also be considered as the dispersed phase. This approach assumes that the dispersed phase is a passive 'species' in the background fluid. It allows separate calculation for these two phases. The influence of the dispersed phase on the main flow and vice versa is through the source terms in the governing equations. Such a method does not require the additional calculation of mixture properties such as density and viscosity. This is convenient and physically sound in the case of an oil-injected compressor where the two phases are fluids of a different type. Although these two flows are usually calculated from the unique density and viscosity of the vapour-liquid mixture, it is more convenient to take account of the values of the vapour and liquid properties separately with concentration as the blending factor between them.

2.2.2 Constitutive Relations

The numerical method contains information about material properties that have to be incorporated into the model. These are used to express the stress tensor $\mathbf{T}$, heat flux $\mathbf{q}_h$ and diffusion flux $\mathbf{q}_{ci}$. Relatively simple assumptions can be made to define values for these in many engineering circumstances. The stress tensor, which represents the viscous rate of transport of momentum and closes equation (2.4), can be defined for Newtonian fluids by Stokes law as:

$$\mathbf{T} = 2\mu \dot{\mathbf{D}} - \frac{2}{3}\mu \operatorname{div} \mathbf{v} \mathbf{I} - p\mathbf{I}, \tag{2.12}$$

where the rate of strain is defined as:

$$\dot{\mathbf{D}} = \frac{1}{2}\left[\operatorname{grad} \mathbf{v} + (\operatorname{grad} \mathbf{v})^T\right]. \tag{2.13}$$

Superscript T represents the transposed tensor. Stokes law gives the relation between the stress and the rate of deformation for Newtonian fluids.

Solid material can be treated as thermo-elastic. For such solids, the constitutive relation that closes equation (2.4) is Hooke's law. It defines the relation between the stress and strain in solids as:

$$\mathbf{T} = 2\eta \mathbf{D} - \lambda \operatorname{div} \mathbf{u} \, \mathbf{I} + (3\lambda + 2\eta)\alpha_t \Delta T \mathbf{I}, \qquad (2.14)$$

where the strain tensor is:

$$\mathbf{D} = \frac{1}{2}\left[\operatorname{grad} \mathbf{u} + (\operatorname{grad} \mathbf{u})^T\right]. \qquad (2.15)$$

Equations (2.12) and (2.14) have the same form. This allows them to be incorporated into a mathematical model and solved by the same method. By this means, the simultaneous calculation of fluid flow and deformation in solids permits the analysis of fluid-solid interaction.

The viscous part of the stress tensor, which appears in equation (2.8), is now fully defined by equations (2.9), (2.12) and (2.13).

The heat flux through the surface boundary $\mathbf{q}_h$ is defined through Fourier's law as the product of the thermal conductivity κ and the temperature gradient.

$$\mathbf{q}_h = \kappa \operatorname{grad} T. \qquad (2.16)$$

The mass flux of the dispersed phase, relative to the mean flow is defined by equation (2.10), by the use of Fick's law:

$$\mathbf{q}_{c_i} = \rho D_i \operatorname{grad} c_i, \qquad (2.17)$$

where D_i is the mass diffusivity of the dispersed phase. In the case of only one fluid dispersed in the main flow, which is the most common case in oil injected screw compressors, equation (2.17) satisfies the overall mass concentration equation:

$$\mathbf{q}_{c_i} = \rho c_i (\mathbf{v}_{c_i} - \mathbf{v}_m). \qquad (2.18)$$

The velocity of the dispersed phase is $\mathbf{v}_{ci}$, while the mass averaged velocity of the mixture is $\mathbf{v}_m = \sum_{i=0}^{N} c_i \mathbf{v}_{c_i}$. Equation (2.18) is satisfied if the fluid is dispersed. Otherwise, it is valid if each fluid satisfies its own equation. On the other hand, Fick's law satisfies equation (2.18) only if $\sum_{i=0}^{N} c_i \mathbf{q}_{c_i} = 0$. This happens only if the diffusion coefficients of all the dispersed fluids have the same value.

Even if all the variables, which define the material properties, are known, the system of equations is still not closed because the pressure p exists in both, the energy equation (2.8) and in the stress tensor (2.12) which forms part of the momentum equation (2.10). An equation of state, which balances the mass equation with thermodynamic properties, usually pressure and temperature, is then required to close the system. This is normally of the form:

14   2 Computational Fluid Dynamics in Screw Machines

$$\rho = \rho(p,T), \quad e = e(\rho,T) \tag{2.19}$$

Equations of state are directly applicable to all engineering fluids and solids, both ideal and real. Common examples are incompressible fluids and solids where ρ=const, or ideal gases where ρ=p/RT. However, real fluids are not rare in screw machinery. In that case, the density of the real gas or mixture can be calculated based on a user specified procedure and later introduced to the model. These equations must have the form of equation (2.19).

2.2.3 Multiphase Flow

In an oil-flooded screw compressor, the dispersed phase in the working fluid is comprised of both the liquid part of the working fluid and the oil. Both components give flow through the machine a multiphase character. There are two different approaches to multiphase flows. One of them is the Eulerian approach where each of two or more phases is contained in its own domain, strictly separated in space from any other but connected with them through a boundary interface. An example of this is an oil tank in which the level of oil is above space occupied by water. If the Eulerian approach is assumed, then a sharp interface exists between the oil and the water and separate numerical meshes can be generated. Both, the oil and the water have to satisfy the governing equations described in 2.2.1.

This is not applicable to two-phase flow within a screw machine. Here, the so-called Eulerian-Lagrangian approach is more appropriate, in which both phases occupy a common working domain without a strict interface between them. In such a case, the background fluid must satisfy the governing equations of mass, energy and momentum, while the dispersed phase should satisfy the governing equation of concentration. Such an approach allows for the dispersed phase to be either a passive or an active component. The dispersed phase in the form of oil or other injected liquid has an important role in the screw compressor working cycle. It is there to cool the fluid, seal the clearance gaps and lubricate the compressor moving parts. The influence of the dispersed phase on the background fluid and vice versa must be incorporated in the governing equations. This is done through source terms in the mass, momentum and energy equations.

The Energy Source

The energy balance of a dispersed phase trapped in the control volume can be written in the following form:

$$\frac{d(m_i h_i)}{dt} = m_i \frac{dh_i}{dt} + h_{iL} \frac{dm_i}{dt} = \dot{Q}_{con} + \dot{Q}_{mass} \tag{2.20}$$

The first term on the right hand side of this equation represents the convective heat flux between the dispersed phase and the background fluid while the second term

2.2 Continuum Model applied to Processes in Screw Machines

represents the heat transfer due to mass interchange between the phases. The last one is significant only for two-phase flow of the same fluid in the working chamber. In the case of an oil-flooded compressor, the convection term has a significant role. In equation (2.20) h_i is the specific enthalpy per unit mass while h_{iL} defines the specific enthalpy of vaporisation. It represents the difference between the specific enthalpies of the liquid and vapour phases, i.e. the dispersed and continuous phases. If the specific heat of oil is constant, then equation (2.20) becomes:

$$\frac{d(m_i T_i)}{dt} = m_o C_{P_o} \frac{dT_o}{dt} + h_L \frac{dm_L}{dt} = \dot{Q}_{con} + \dot{Q}_{mass}, \qquad (2.21)$$

There are two approaches to allow for convective heat transport between the dispersed phase and the background fluid.

The first approach assumes that the dispersed phase is completely dissolved in the background fluid. That means that the droplet size of the dispersed phase is very small, ie. $d_o \to 0$. In that case, an ideal process of heat transfer can be assumed where the temperature of the dispersed phase is assumed to be equal to the temperature of the background fluid $T=T_o$. Heat exchange between the phases is then calculated from the temperature difference of the continuous phase, at two consecutive instants of time, multiplied by the mass and specific heat of the dispersed phase as:

$$\dot{Q}_{con} = m_o C_{P_o} \frac{dT}{dt} \approx m_o C_{P_o} \frac{T^k - T^{k-1}}{\delta t}, \qquad (2.22)$$

where T^k is the temperature in the current time step and T^{k-1} is the temperature from the previous time step or iteration. δt is the time step. If the time step is small then this equation has the exact differential form of convective heat transfer. The assumption of an infinitesimally small droplet size is not completely correct but analysis of the influence of oil on screw compressor process performance by *Stosic et al* (1992) showed that a change in droplet size from 0 to 10 μm, does not affect the oil and consequently the gas temperature very much. Therefore, it is accurate enough to calculate the heat exchanged between the continuous and dispersed phases by means of equation (2.22).

When necessary, another approach can be used to calculate the convective heat transfer term in equation (2.21). It should be applied whenever the temperatures of the continuous and dispersed phases cannot be considered to be equal. It is then assumed that the dispersed phase, contained in the control volume, consists of spherical droplets with a Sauter mean diameter defined as:

$$d_o = 0.0092 \left(\frac{\rho_o \sigma_o}{|\mathbf{v}|} \right) \left(1 + \frac{1}{c_o} \right)^{0.7} \qquad (2.23)$$

where σ_o is the surface tension and $|\mathbf{v}|$ is the absolute value of the fluid velocity. The convective heat flux then becomes:

$$\dot{Q}_{con} = \pi d_o \kappa \, \text{Nu} (T - T_o), \tag{2.24}$$

where, T and T_o are temperatures of the continuous and dispersed phases respectively and the Nusselt number is given by:

$$\text{Nu} = 2 + 0.6 \, \text{Re}^{0.5} \, \text{Pr}^{0.33}. \tag{2.25}$$

In the previous equation the Prandtl number is defined as:

$$\text{Pr} = \frac{\mu C_p}{\kappa} \tag{2.26}$$

Reynolds number is:

$$\text{Re} = \frac{\rho |\mathbf{v}_o - \mathbf{v}| d_o}{\mu}, \tag{2.27}$$

The velocity of the dispersed phase is $|\mathbf{v}_o|$. There are again two possible approaches. The first is to assume the velocity of the dispersed phase to be the same as the velocity of the continuous phase. In this case $\mathbf{v}=\mathbf{v}_o$. This can be assumed if the size of the droplet defined by equation (2.23) is small, e.g. less then 20 μm. If this is not the case, a different approach has to be applied and the velocity vector of the dispersed phase has to be calculated by another procedure. Whatever the velocity, the temperature of the dispersed phase is derived from the balance of two equations that define heat transfer namely: (2.22), which represents the amount of heat taken in by the dispersed phase, and (2.24) which defines amount of heat given out by the continuous phase. This can be written as:

$$m_o C_{p_o} \frac{T_o^k - T_o^{k-1}}{\delta t} = \pi d_o \kappa \, \text{Nu} (T - T_o^k), \tag{2.28}$$

where T_o^k and T_o^{k-1} are temperatures of the dispersed phase in the two consecutive time steps.

When equation (2.25) is applied, the temperature of the dispersed phase becomes:

$$T_o = T_o^k = \frac{T + k_t T_o^{k-1}}{1 + k_t}, \tag{2.29}$$

where the time constant k_t is defined as:

$$k_t = \frac{m_o C_{p_o}}{\pi d_o \, \mathrm{Nu} \, \kappa \, \delta t} \qquad (2.30)$$

In all the previous equations, the mass of the dispersed phase in the control volume is calculated from the mass concentration c_o defined from equation (2.10) as:

$$m_o = \rho_o V_o = \frac{\rho_o \rho V c_o}{\rho(1-c_o) + \rho_o c_o} \qquad (2.31)$$

where ρ_o and ρ are the densities of the dispersed and the continuous phases respectively. Both densities are calculated from the equation of state (2.19).

The last term in equation (2.21) represents heat transfer due to mass transfer between the phases. It is significant when a real fluid evaporates or condenses in the machine. It can be expressed as:

$$\dot{Q}_{mass} = h_L \frac{dm_L}{dt} \approx h_L \frac{\Delta m_L}{\delta t} = h_L \frac{m_L - m_L^s}{\delta t} = h_L \dot{m}_L, \qquad (2.32)$$

where Δm_L is the mass exchanged between the liquid dispersed phase and the continuous phase. It is defined as the difference between the mass of the continuous phase in the control volume, calculated from the mass balance equation (2.2), and the mass of the continuous phase, calculated by the equation of state (2.19), at the pressure obtained from the governing equations and the saturation temperature at the same pressure.

The latent heat of evaporation h_L is the difference between the specific enthalpy of the liquid h_l and the specific enthalpy of the vapour h_v at saturation pressure p, which is calculated from the model:

$$h_L = h_v - h_l \qquad (2.33)$$

Since these two specific enthalpies at present are not known they should be calculated together with other properties of the real fluid.

The heat fluxes calculated from equations (2.24) and (2.32) represent the source terms in the energy equation (2.8).

The Mass Source

The mass of the dispersed phase changes in two-phase flow because of evaporation or condensation in the control volume. The amount of mass exchanged between the two phases is defined as:

$$\dot{m}_L = \frac{dm_L}{dt} \approx \frac{\Delta m_L}{\delta t} = \frac{m_L - m_L^s}{\delta t}. \qquad (2.34)$$

In practice, if the control volume is assumed constant during one time step, the pressure and temperature are calculated from the governing equations together with the mass of the continuous phase. The mass concentration of the dispersed phase is calculated from equation (2.10). This procedure defines the mass of the dispersed phase m_o through equation (2.31). If two-phase flow exists in the control volume, the temperature of the mixture is the saturation temperature for the calculated pressure. If the temperature calculated by the model does not satisfy this condition, mass must be exchanged between the dispersed and continuous phases to establish equilibrium. The exchanged mass transfers the heat of evaporation between the phases until equilibrium is established. The heat of evaporation is calculated from the balance equation of the heat exchanged between the phases and the heat required to adjust the temperature of the mixture to the saturation temperature for the pressure calculated in the control volume:

$$\Delta m_L \cdot h_L = m \cdot C_{pm} \cdot (T - T_{sat}) \qquad (2.35)$$

This mass becomes either a source or a sink in the mass equation for the continuous phase. Also, it is subtracted from the concentration of the dispersed phase through the source term in equation (2.10).

The Momentum Source

The equation of motion for an individual droplet of the dispersed phase in a positive displacement machine is given in the form of an ordinary differential equation based on Newton's second law:

$$\frac{d(m_o \mathbf{v}_o)}{dt} = \mathbf{f}_{drag} + \mathbf{f}_{pres} + \mathbf{f}_{body} + \mathbf{f}_{am}, \qquad (2.36)$$

where, often, the pressure forces, $\mathbf{f}_{pres}$, body forces, $\mathbf{f}_{body,}$ and apparent mass forces, $\mathbf{f}_{am,}$ can be neglected. The interphase drag force $\mathbf{f}_{drag}$ is:

$$\mathbf{f}_{drag} = -\frac{1}{2} \rho A_o C_{drag} |\mathbf{v}_o - \mathbf{v}| (\mathbf{v}_o - \mathbf{v}), \qquad (2.37)$$

where $A_o = d_o^2 \pi / 4$ is the surface of the dispersed phase particle with Sauter mean diameter d_o, $\mathbf{v}_o$ is the velocity of the dispersed phase in the control volume and C_{drag} is the drag coefficient which, in case of a Newtonian fluid, depends only on the Reynolds number defined by equation (2.27). When applicable, equation (2.36) is used to calculate the velocity of the dispersed phase while equation (2.27) gives a source term for the momentum equation. If one assumes ideal heat transfer with a particle of size 0, then the drag force is also 0.

2.2.4 Equation of State of Real Fluids

Refrigerating and air conditioning and process compressors, as well as process gas compressors operate with real fluids i.e. where the assumption of ideal gas relationships is too gross. In such a case, complex functions are required to describe the fluid property changes. Commercial software packages are available today for the calculation of real fluids. Most of these packages are impractical for use in CFD because of the large number of calculations required to obtain the required thermodynamic properties. However, often users develop property software for their own requirements which gives good agreement over the required range of operating conditions.

In the case of two-phase flow, the required thermodynamic properties are: saturation temperature, density of the mixture, specific heat of the mixture, latent heat of evaporation and C_p. The latter is a constant that appears in the mass flux correction in the coupling procedure of the mass equation and equation of state. It defines the rate of change of density with change in pressure to correct the pressure-velocity coupling procedure. It is expressed as:

$$C_\rho = \left(\frac{\partial \rho}{\partial p}\right)_T, \tag{2.38}$$

for constant temperature in one iteration. In the case of an ideal gas, the value of this constant is derived directly from the equation of state $p/\rho = RT$ as:

$$C_\rho = \left(\frac{\partial \rho}{\partial p}\right)_T = \frac{\rho}{p} = \frac{1}{RT}. \tag{2.39}$$

However, for a real fluid, the equation of state is expressed as:

$$\frac{p}{\rho} = z\,RT = z(p,T)\,RT, \tag{2.40}$$

where z is the compressibility factor. This is generally a non-linear function of pressure and temperature. There are approximations derived for this factor and it is assumed here to be a linear function of the working pressure:

$$z = p\,B_1 + B_2, \tag{2.41}$$

where B_1 and B_2 are constants with different values for each fluid. For the ideal gas $B_1=0$ and $B_2=1$. The compressibility factor is approximated such that coefficients B_1 and B_2 are calculated from measured thermodynamic properties of saturated vapour at pressures of 1 and 20 bar. Screw machines usually operate within this range of working pressures regardless of their application and this approximation does not involve a large error in the estimation of thermodynamic properties. It leads to a maximum error of approximately 2% at 10 bar. This is sufficiently accurate, but outside this range different coefficients need to be used.

If the compressibility factor at the working pressure and temperature is known, then the density of the vapour or gas is derived from:

$$\rho_v = \frac{p}{zRT}, \qquad (2.42)$$

It can be assumed that liquids in screw machines used for lubrication and generated as a result of the condensation process, are incompressible at the machine pressures. This means that the density of the liquid is constant:

$$\rho_l = const, \qquad (2.43)$$

The density of a liquid-vapour mixture in the saturated domain can be written as:

$$\rho = \frac{1}{\frac{1-c_2}{\rho_v} + \frac{c_2}{\rho_l}} = \frac{p}{(1-c_2)zRT + \frac{c_2}{\rho_l}p}, \qquad (2.44)$$

To obtain an equation for C_ρ, the temperature is regarded as constant within the iteration. The first derivative of equation (2.44) gives:

$$\left(\frac{d\rho}{dp}\right)_T = \frac{1-(1-c_2)\rho RT\frac{dz}{dp} - \frac{c_2}{\rho_l}\rho}{(1-c_2)zRT + \frac{c_2}{\rho_l}p}. \qquad (2.45)$$

The derivative in the second term on the right side represents the change of compressibility factor with pressure. It follows from equation (2.41) that this derivative is constant and has the value B_1. Introducing that feature into the previous equation, the coefficient C_ρ can finally be obtained as:

$$C_p = \left(\frac{d\rho}{dp}\right)_T = \frac{1-(1-c_2)\rho RT\, B_1 - \dfrac{c_2}{\rho_l}\rho}{(1-c_2)zRT + \dfrac{c_2}{\rho_l}p}. \qquad (2.46)$$

If as a consequence of the pressure and temperature in the control volume, the working fluid is liquid, which gives $c_2 = 1$, the coefficient C_p reads zero and its value is not a function of pressure. If only vapour or real gas occupy the working chamber, $c_2 = 0$ and equation (2.46) becomes:

$$C_p = \left(\frac{d\rho}{dp}\right)_T = \frac{1}{zRT} - \frac{\rho_v B_1}{z}. \qquad (2.47)$$

If the fluid is ideal, then B_1 becomes zero and z tends to one. In that case equation (2.46) becomes the same as equation (2.39). However, if the fluid is real, B_1 becomes slightly negative and z tends to values less then 1. This means that the second term becomes positive and it contributes to the value of the first term. The value of that term increases with the change of the 'fluid reality', which is expressed through constants B_1 and B_2 in (2.41). This term becomes significant in comparison with the first term if the fluid is real. In the case of ammonia, for example, at a pressure of 5 bar the first term has a value of $6.7 \cdot 10^{-6}$ while the value of the second term is $2.9 \cdot 10^{-2}$. The coefficient C_p derived from equation (2.47) is later used for the calculation of pressure in the pressure-velocity-density coupling procedure.

Other thermodynamic properties are not directly derived from the equation of state but, as a consequence of the fluid behaviour, these are calculated from thermodynamic properties of both the liquid and vapour phases.

The saturation temperature is calculated from a modified version of Antoine's equation, which is in its original form expressed as:

$$\log p = A_1 - \frac{A_2}{t + A_3} \qquad (2.48)$$

which is an explicit expression for saturation pressure as a function of temperature, *Walas* (1984). Constants A_1, A_2 and A_3 vary for different fluids and they are obtained from experimental results. The value of the coefficient A_3 is usually small and in many cases can be neglected. In that case, the equation explicitly gives saturation temperature in terms of pressure as:

$$T_{sat} = \frac{A_2}{A_1 - \log p} - A_3 \qquad (2.49)$$

The saturation temperature calculated from the previous equation is used in equation (2.35) to estimate the mass exchanged during evaporation/condensation. That equation gives the mass which transfers the latent heat of evaporation from one phase to the other. The latent heat of evaporation is calculated for the saturation pressure by means of the Clapeyron equation. This is expressed as:

$$h_L = T_{sat}\, v_{lv} \left(\frac{dp}{dT}\right)_{sat}, \qquad (2.50)$$

where v_{lv} is the difference between the vapour and liquid specific volumes. Typically, more about equation (2.50) can be found in *Cengel and Boles* (1989).

The specific heat at constant pressure is a fluid property needed to calculate the specific enthalpy of the mixture. The specific heat of the mixture C_{pm} is the weighted sum of the specific heats of vapour C_{pv} and liquid C_{pl} for constant pressure, ie:

$$C_{pm} = (1-c_2)C_{pv} + c_2 C_{pl} \qquad (2.51)$$

The specific heat of vapour can be calculated from the following equation:

$$C_{pv} = D_0 + D_1 T + D_2 T^2 + D_3 T^3 \qquad (2.52)$$

where D_0, D_1, D_2 and D_3 are constants which vary for different fluids. Their values can be found in *Sonntag and Borgnakke* (2001). If the specific heat of liquid at constant pressure is assumed constant, which is reasonably accurate over a limited temperature range, then equation (2.51) gives the specific heat of the mixture. Even if the concentration of the liquid phase in the working chamber is equal to zero, this equation can be used to express the specific heat of the working fluid, which in this case is vapour.

By use of the equations derived in this Section, the properties of real fluids, which are liquid, vapour or their mixture, are completely described. The procedure is fast and efficient for calculation in the numerical CFD solver, because all equations are analytical and the variables are derived explicitly from the pressure, the value of which is obtained from the mass-velocity-pressure coupling procedure. The procedure is equally applicable to ideal gases, and incompressible fluids. The coefficient C_p calculated from the equation (2.46) is used in the next iteration as a source term.

2.2.5 Turbulent Flow

Turbulent flows are well described by the governing differential equations presented in section 2.2. However, their direct numerical simulation requires a mesh with spacing smaller than the length scale of the smallest turbulent eddies, at which the energy is transformed to heat, and time steps smaller than the smallest time scale of the turbulent fluctuations. Some calculations show that the average length scale of the smallest eddies in positive displacement machines is of the order of $10 \mu m$ while their time scale is of the order of a couple of milliseconds. This requires computer resources, which are not yet available.

Alternatives are either large eddy simulation, in which only the largest unsteady motions are resolved and the rest is modelled, or a solution to the Reynolds averaged Navier-Stokes equations where all turbulent effects on the mean flow are modelled as functions of mean fluid flow quantities.

The Reynolds averaged Navier-Stokes equations (RANS) are obtained by using a statistical description of the turbulent motion formulated in terms of averaged flow quantities. Many such models of turbulence are developed to date, which are suitable for different fluid flow situations. Only two of them are described in some details in Appendix A. These are the Zero-Equation model and the Standard k-ε two-equation model. More details on turbulence phenomena can be found for example in *Wilcox* (1993).

2.2.6 Pressure Calculation

The pressure in the source term of the fluid momentum equation is unknown because it does not appear explicitly in the continuity equation. This constraint is satisfied only if the pressure field is adjusted to the resulting fluid flow. The method of calculation of the pressure and pressure gradient fields consists now of three steps. The first one is to obtain the velocity field from the momentum equation regardless of whether the continuity equation is satisfied or not. The second is a predictor stage in which a pressure correction is calculated to satisfy the continuity equation and the third one is a corrector stage in which new values of the velocity, pressure and density fields are calculated. The method is known as a SIMPLE algorithm and is described in Appendix C.

2.2.7 Boundary Conditions

Special treatment of boundaries is introduced due to the compressor communicating with its surroundings through small receivers at suction and discharge and also through oil injection. The common practice is to keep the pressure in these receivers constant. Therefore, an appropriate amount of mass and energy is added or subtracted from these receivers.

Wall Boundaries

There are two types of walls applied to a screw compressor; moving walls, if they bound the domain on the compressor rotors, and stationary walls in other places. Boundary conditions on these walls are explicitly given for all equations in the numerical model. In the case of turbulent flow, dependent variables vary steeply near the solid boundaries and a method, which can model near wall effects, is used. If the flow is laminar, then the dependent variable is either known on the boundary, or its flux is given on the boundary. The walls are treated as 'no-slip walls', which is the case when viscous fluid sticks to the boundary wall.

Boundary conditions for the momentum equation are given through the known velocities on the wall. For the rotor walls, the velocities are calculated from the given rotational speed n of the male rotor as:

$$\omega_1 = \frac{2n\pi}{60}; \qquad \mathbf{v}_{1i} = \mathbf{r}_{1i}\,\omega_1$$

$$\omega_2 = -\omega_1 \frac{z_1}{z_2}; \qquad \mathbf{v}_{2i} = \mathbf{r}_{2i}\,\omega_2$$

(2.53)

Subscript *1* indicates the male rotor while the value *2* is related to the female rotor. z_1 and z_2 are the number of lobes on the rotors, $\mathbf{r}_{1i}$ and $\mathbf{r}_{2i}$ are radius vectors of the boundary points on the male and female rotors respectively in an absolute coordinate system. ω_1 and ω_2 are the angular velocities on the male and female rotors respectively. For all stationary walls, the wall velocity is equal to zero.

More details of the equations, which incorporate wall boundaries to the mathematical model of the screw compressor process, are given in B.

Constant Pressure in the Inlet, Outlet and Oil Receivers

Even if the compressor cycle can be considered steady, this is true only if it is averaged in time over a period in which a compressor completes a number of cycles. However, within one cycle, the compressor system is always in a state of transition. Such a transition is caused by rotation of the rotors, which moves the corresponding part of the numerical mesh. That movement is defined by the angular velocity ω_r. Movement of the computational domain causes change in its volume, which further causes pressure change within it. The pressure difference between the cells causes fluid to flow through the machine. Contrary to the rotor domains, other parts of the compressor such as the inlet and outlet ports and receivers maintain a constant volume. The fluid flow induced between the rotors inevitably leads to change of the pressure in the parts which keep a constant volume. In a real compressor, such a situation causes additional fluid to flow into or out of the chambers, keeping the pressure constant. This process is simulated in the numerical procedure.

The first possibility is to apply standard inlet and outlet boundaries. However, in that case, either the inlet velocity or the mass flux should be prescribed in advance, which is extremely difficult. The compressor flow depends on the rotor speed and varies considerably during the cycle. Additionally, reverse flow can exist at the outlet boundary if it is not far enough from the discharge port. That situation is not allowed with the standard boundary conditions. Therefore, these boundary conditions are not adequate for a screw compressor.

The other possibility is to apply pressure boundaries at the inlet and outlet. In the standard pressure boundary condition a prescribed pressure on the boundary is combined with the following boundary condition:

$$\left(\frac{\partial \mathbf{v}}{\partial n}\right)_B = 0 \tag{2.54}$$

to obtain boundary velocities $\mathbf{v}_B(\mathbf{r}_b,t)$. Other treatments are necessary in the case of supersonic and subsonic flows. If the outlet flow is supersonic, then both the pressure and the velocity should be obtained by extrapolating from the upstream region. It is obvious that the pressure boundary conditions are similar to the inlet or outlet boundaries, firstly because they couple pressure and velocity directly and secondly because for all equations, apart from the momentum equation, the boundary properties are calculated from the velocity. This procedure may cause instability in the compressor cycle especially when the flow changes its direction at the boundary.

In opposition to both types of boundary condition mentioned above, application of the boundary domain, in which an amount of mass is added or subtracted to maintain constant pressure, is natural and gives a stable and relatively fast solution.

Starting from the equation of state for a real fluid (2.40) for constant instant temperature and density of fluid in a receiver of volume V, or in an individual numerical cell of volume V_i, the following equation can be derived:

$$\dot{V}_{add} = \frac{\partial V_i}{\delta t} = \frac{V_i}{p_i}\frac{\partial p_i}{\delta t} \tag{2.55}$$

It gives the relation between the volume change and the pressure change. The value of V_i is the volume flow that corresponds to the change in pressure ∂p_i during time ∂t. As the density is assumed constant, then the mass flux, which corresponds to the pressure change, is:

$$\dot{m}_{add} = \rho_i \frac{V_i}{p_i}\frac{\partial p_i}{\partial t} \tag{2.56}$$

This is the amount of mass, which must be added or subtracted to a receiver of constant volume V or to an individual numerical cell placed in the considered receiver to maintain constant pressure.

The amount of mass calculated from (2.56) represents a mass source in the pressure correction procedure explained in Section 2.2.6. It will maintain constant pressure in the considered cell and the momentum equation correction would result in a new velocity in the cell.

The energy equation is corrected in order to keep the system in balance. It is done through the volume source in the governing equation of energy:

$$\dot{q}_{add} = m_i C_{pi} \frac{\partial T_i}{\partial t} = m_i C_{pi} \frac{T_i - T_r}{\partial t} . \qquad (2.57)$$

In the last term of equation (0.57) T_i is the temperature calculated in the cell, T_r is the specified temperature which has to be maintained and ∂t is the time step.

When an amount of mass of dispersed phase is added to the selected set of cells, the equation for species also has to be updated. The concentration of the dispersed phase can be known, or prescribed, in some domains while in others it has to be estimated. For example, the concentration of oil in the oil injection port always has a value which is close to 1. Similarly, the concentration of liquid in the inlet port of a two-phase expander is defined by the pressure and quality of the mixture. However, there are some compressor domains where the value of concentration is not known but the pressure has to be maintained constant. In that case, the value of concentration must be extrapolated from the neighbouring domain.

When the concentration is known, then its value should be kept as close as possible to the prescribed value. The mass of the dispersed phase carried by the continuous phase is calculated by equation (2.31) in which the concentration c_o is substituted by the prescribed value c_p. The last term in the transport equation for the concentration of the dispersed phase (2.10), is the volume source term, which is expressed as:

$$\int_V S_{c_i} dV = S_{c_i} V_i = \rho \frac{\partial c_i}{\partial t} V_i = \dot{m}_i . \qquad (2.58)$$

This volume source, when integrated over the cell volume, is the amount of mass of the dispersed phase added to or subtracted from the mass of the numerical cell. If the concentration of the dispersed phase has to be maintained constant, a correction to the equation of species has to be added through the volume source. The source term in equation (2.10) is the mass flux of the dispersed phase. Its calculation is based on the desired concentration of the dispersed phase. Equation (2.31) is used for that and c_o is replaced by the desired concentration in the cell. In such a situation, the volume source in the oil concentration equation becomes:

$$S_{c_o} = \int_V S_{c_o} dV = m_{o_{add}} = \frac{\rho_o \rho_m V c_{o_{cons}}}{\rho_m (1 - c_{o_{cons}}) + \rho_o c_{o_{cons}}} \qquad (2.59)$$

In the case of two-phase flow with evaporation or condensation, the equation for the concentration of the liquid phase of the working fluid has to be updated through its volume source term. This term is calculated from equation (2.34) as:

$$S_{c_l} = \int_V S_{c_l} dV = -\dot{m}_L \approx -\frac{\Delta m_L}{\delta t} \tag{2.60}$$

Other equations, like these for the turbulence model, do not need to be updated for this case.

2.3 Finite Volume Discretisation

2.3.1 Introduction

The finite volume method is employed to solve fluid flow equations. It can also be applied to solid body stress analysis, independently or when coupled with fluid flow. The method is fully implicit and can accommodate both structured and unstructured moving grids with cells of arbitrary topology. Although the procedure is described here for fluid flow in screw compressors, it is general and can be used for any physical problem which is described by the given equation set.

A segregated approach is used to solve the resulting set of coupled non-linear system of algebraic equations. The equations are solved by an iterative conjugate gradient solver which retains the sparsity of a coefficient matrix, thus achieving the efficient use of computer resources.

If an appropriate constitutive relation is applied to each conservation law, namely mass, momentum and energy balance, a closed set of M equations is obtained for each numerical cell in a particular time step. The number of equations M depends on the problem that has to be solved. 7 equations are required for a screw compressor, including two-phase flow with oil injection. All the conservation equations can be conveniently written in the form of the following generic transport equation:

$$\frac{d}{dt}\int_V \rho\phi dV + \int_S \rho\phi(\mathbf{v}-\mathbf{v}_s)\cdot d\mathbf{s} = \int_S \Gamma_\phi \mathrm{grad}\,\phi \cdot d\mathbf{s} + \int_S \mathbf{q}_{\phi S}\cdot d\mathbf{s} + \int_V q_{\phi V}\cdot dV \tag{2.61}$$

The continuity equation is combined with the momentum equation to obtain an equation for pressure or pressure correction. The meaning of symbols used in this equation is given in the nomenclature. The diffusive flux and sources are given in

Table 2-1 for each property ϕ.

28 2 Computational Fluid Dynamics in Screw Machines

Table 2-1 Terms in the generic transport equation (2.61)

Equation	ϕ	Γ_ϕ	$\mathbf{q}_{\phi S}$	$q_{\phi V}$
Continuity	1	0	0	0
Fluid Momentum	v_i	μ_{eff}	$\left[\mu_{\text{eff}}(\text{grad}\,\mathbf{v})^T - \left(\frac{2}{3}\mu_{\text{eff}}\text{div}\,\mathbf{v} + p\right)\mathbf{I}\right]\cdot\mathbf{i}_i$	$f_{b,i}$
Solid Momentum	$\dfrac{\partial u_i}{\partial t}$	η	$\left[\eta(\text{grad}\,\mathbf{u})^T + (\lambda\text{div}\,\mathbf{u} - 3K\alpha\Delta T)\mathbf{I}\right]\cdot\mathbf{i}_i$	$f_{b,i}$
Energy	e	$\dfrac{k}{\partial e/\partial T}+\dfrac{\mu_t}{\sigma_T}$	$-\dfrac{k}{\partial e/\partial T}\dfrac{\partial e}{\partial p}\cdot\text{grad}\,p$	$\mathbf{T}:\text{grad}\,\mathbf{v}+h$
Concentration	c_i	$\rho D_{i,\text{eff}}$	0	s_{ci}
Space	$\dfrac{1}{\rho}$	0	0	0
Turbulent kinetic energy	k	$\mu+\dfrac{\mu_t}{\sigma_k}$	0	$P-\rho\varepsilon$
Dissipation	ε	$\mu+\dfrac{\mu_t}{\sigma_\varepsilon}$	0	$C_1 P\dfrac{\varepsilon}{k}-C_2\rho\dfrac{\varepsilon^2}{k}-C_3\rho\varepsilon\,\text{div}\,\mathbf{v}$

The ability of expressing all transport equations in the form of a prototype equation (2.61) facilitates the discretisation procedure, which together with the appro5priate initial and boundary conditions, forms the mathematical model of continuum mechanics problems.

The Finite Volume Method (FVM) is used to discretise the governing equations. All dependent variables are stored in a collocated variable arrangement, which requires only one set of control volumes to be generated. This enables eventual implementation of the multigrid method and local grid refinement.

Equation (2.61) can be written for a control volume in a Cartesian coordinate frame, as presented in Figure 2-2. This equation is still general and exact:

$$\frac{d}{dt}\int_V \rho\phi\,dV + \sum_{j=1}^{n_f}\int_{S_j}\rho\phi(\mathbf{v}-\mathbf{v}_s)\cdot d\mathbf{s} = \sum_{j=1}^{n_f}\int_{S_j}\Gamma_\phi\,\text{grad}\,\phi\cdot d\mathbf{s}+\left(\sum_{j=1}^{n_f}\int_{S_j}\mathbf{q}_{\phi S}\cdot d\mathbf{s}+\int_V q_{\phi V}\cdot dV\right) \quad (2.62)$$

It consists of four terms which describe the effects of rate of change with time, convection, diffusion and source respectively. For each cell, all quantities are then written in the form of equation (2.62) and set up as a system of $n*m$ partial differential equations. Each cell acts as a control volume, the total number of which is n, while the number of unknowns for each cell is m. These are all transferred to a system of $n*m$ algebraic equations in order to be solved numerically.

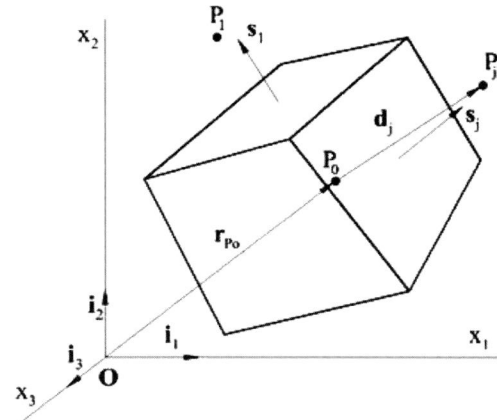

Figure 2-2 Notation applied to a hexahedral control volume

Therefore, the surface and volume integrals, appearing in the equations are replaced by quadrature approximations, the spatial derivatives are replaced by an interpolation function and the time integration scheme is selected and after that the control volume surface velocities **v**s are determined.

2.3.2 Space Discretisation

In this work, space is discretised by an unstructured mesh with polyhedral control volumes with an arbitrary number of faces. However, hexahedra are used wherever possible, which facilitates the local grid refinement. In some cases this may be essential for accuracy. The spatial discretisation of a screw compressor working domain is presented in a separate chapter.

2.3.3 Time Discretisation

The time interval of interest is divided into an arbitrary number of subinterval time steps, which are not necessarily of the same duration. However, the procedure used for mesh movement requires the time steps in the simulation procedure of the screw compressor working cycle to be constant. It is aimed that all variables at the start and end of a calculation cycle are equal. The calculation cycle is represented by rotation of the male rotor either for a full revolution or only for one lobe rotation. The constant time step, however, is not given arbitrarily. It depends on the chosen number of rotational steps within the tooth span angle on the male rotor and the speed of rotation n. The angular speed of the male rotor is:

$$\omega_1 = 2\pi \frac{n}{60} \quad [rad/s]. \tag{2.63}$$

If the compressor rotates at constant speed, the unit angle is:

$$\delta\varphi = \frac{\varphi}{n_{ang}} \qquad (2.64)$$

where φ is the interlobe angle of the male rotor and n_{ang} is number of divisions of that angle for the rotation of one full interlobe. The time step is then defined as

$$\delta t = \frac{\delta\varphi}{\omega_1} \qquad (2.65)$$

Although the time step in the majority of calculations of screw compressor performance is constant, if a transient state has to be calculated, the time step changes during marching in time. This is especially the case for compressor start up and shut down procedures. In these cases, the time step within two consequent rotations of the compressor rotors depends on the first time derivative of the compressor speed. Again, it is necessary to calculate the angular velocity for each time step and consequently to update the time step.

2.3.4 Discretisation of Equations

Discretisation Principles

The result of discretisation of the prototype equation (2.62) is a system of algebraic equations. Surface and volume integrals are replaced by quadrature approximations, spatial derivatives are replaced by some interpolation function and either a two-times-level or a three-times-level integration scheme in time is selected. These procedures are explained in detail in Appendix C.

Boundary and Initial Conditions

The boundary conditions on the cell faces which coincide with the solution domain boundary are applied prior to solution of the algebraic equations. All compressor solid parts are no-slip walls with either, a known temperature or a temperature approximated, in advance, through a known procedure. Therefore a cell face flux ϕ_j^* represents the boundary flux ϕ_B for all equations in the boundary cells. In such a case, the mass flux in the momentum equation at the boundary is zero, the heat flux through the boundary for the energy equation is calculated from the wall temperature and the thermal conductivity in the near wall region, while the concentration flux is zero. Diffusive fluxes are also replaced with their boundary values.

Screw compressor flow simulation is transient, which requires initial conditions to be prescribed for the dependent variables in each control volume. A close match of these is important for quick solution convergence.

The initial values of the velocities in the momentum equation are set to zero in all cells within the working chamber. The initial pressures prescribed for the cells in the inlet and outlet receiver are the inlet and outlet pressure. For all other cells the initial values are calculated by linear interpolation between these values with respect to the relative distance in the z direction as:

$$p_i^0 = p_{inl}^0 + \frac{z_i}{L}\left(p_{out}^0 - p_{inl}^0\right). \tag{2.66}$$

z_i is the cell centre distance starting from the coordinate origin, while L is the overall compressor length. This simple method to prescribe initial values often gives a consistent final solution within 4 to 5 compressor cycles. The initial temperature is calculated in the same manner as the pressure by linear interpolation between prescribed inlet and outlet temperatures. The density is then calculated according to equation (2.44) Concentration is also interpolated between the prescribed values at the inlet ρ_{inl}^0 and outlet ρ_{out}^0 of the compressor similarly to the other variables. The initial values of kinetic energy and its dissipation rate are set as zero throughout the domain.

If implicit time integration is employed, these prescribed values at time t_0 are sufficient for the calculation. If, however, the three time level implicit scheme is used, values at the time $t_{-1} = t_0 - \delta t_0$ must be given. They are set at the same value as those at time t_0.

Derived System of Algebraic Equations

If discretisation methods and boundary conditions are implemented in the prototype equation (2.62) for all control volumes then the derived algebraic equation has the same form for all variables:

$$a_{\phi 0}\,\phi_{P_0} - \sum_{j=1}^{n_j} a_{\phi j}\,\phi_{P_j} = b_\phi \tag{2.67}$$

where index 0 determines the control volume in which the variable is calculated and index j defines the neighbouring cells. Symbol n_i represents a number of internal cell faces between the calculating cell and its neighbouring cells. The right hand side contains all terms for which the variables are known from either the previous iteration or the time step. All the coefficients, central $a_{\phi 0}$, neighbouring $a_{\phi j}$ and right hand side b_ϕ, are treated explicitly using a deferred correction approach to increase computational efficiency.

$$a_{\phi j} = \Gamma_{\phi j} \frac{\mathbf{s}_j \cdot \mathbf{s}_j}{\mathbf{d}_j \cdot \mathbf{s}_j} - \min(\dot{m}_j, 0),$$

$$a_{\phi 0} = \sum_{j=1}^{n_f} a_{\phi j} + \frac{(\rho V)_{P_0}^{m-1}}{\delta t_m}, \qquad (2.68)$$

$$b_{\phi} = \sum_{j=1}^{n_f} \Gamma_{\phi j} \left((\operatorname{grad} \phi)_j \cdot \mathbf{s}_j - \overline{\operatorname{grad} \phi} \cdot \mathbf{d}_j \frac{\mathbf{s}_j \cdot \mathbf{s}_j}{\mathbf{d}_j \cdot \mathbf{s}_j} \right) -$$

$$\sum_{j=1}^{n_f} \frac{\gamma_{\phi}}{2} \dot{m}_j \left((\mathbf{r}_j - \mathbf{r}_{P_0}) \cdot (\operatorname{grad} \phi)_{P_0} + (\mathbf{r}_j - \mathbf{r}_{P_0}) \cdot (\operatorname{grad} \phi)_{P_j} + (\phi_{P_j} - \phi_{P_0}) \frac{\dot{m}_j}{\operatorname{abs}(\dot{m}_j)} \right) +$$

$$Q_{\phi S} + Q_{\phi V} + \sum_{B=1}^{n_B} a_{\phi B} \phi_B + \frac{(\rho V \phi)_{P_0}^{m-1}}{\delta t_m}.$$

$\mathbf{d}_j$ is a distance vector. It is effective if the mesh is non-orthogonal and it is then used to correct the cell face value. It is defined as the normal distance between the line connecting two neighbouring cell points and the cell face centre. n_B is the number of boundary faces surrounding the cell P_0. The coefficient $a_{\phi B}$ for the centre point at the boundary cell face is calculated similarly to the neighbouring coefficient $a_{\phi j}$, assuming the distance between the boundary point and the centre of the cell.

2.4 Solution of a Coupled System of Nonlinear Equations

Equations in the form of (2.67) are obtained for each dependent variable like velocity, pressure, temperature and concentration at all points of the computational domain. As a consequence of convective transport and because of other flow characteristics, the equations are non-linear and coupled. In order to be solved, they are linearized and decoupled. The segregate iterative algorithm is adopted.

Coefficients $a_{\phi j}$ and source terms b_{ϕ} are known in advance from the previous iteration or time step. As a result, a system of linear algebraic equations is obtained for each dependent variable. This can be written in matrix notation as:

$$A_{\phi} \boldsymbol{\phi} = \mathbf{b}_{\phi} \qquad (2.69)$$

Here A_{ϕ} is an $N \times N$ matrix, the vector $\boldsymbol{\phi}$ contains values of the dependent variables ϕ at N nodal points in the CV centres and $\mathbf{b}_{\phi}$ is the source vector. The resulting matrix A_{ϕ} obtained by the discretisation method is sparse, with the number of non-zero elements in each row equal to the number of nearest neighbours plus one, $n_i +1$. The matrix is symmetric only for the momentum equation of an elastic solid body and the pressure correction for incompressible fluids. The matrix is diagonally dominant $A_{\phi_0} \geq \sum_{j=1}^{n_i} a_{\phi j}$. This allows solution of the equation system

(2.69) by a number of iterative methods resulting in reasonable computer memory requirements. The conjugate gradient method (CG) is used when the matrix is symmetric and the preconditioned conjugate gradient stable method (CGSTAB) is used when the matrix is asymmetric.

Equation (2.69) is then solved in sequence for each dependent variable. There is no need to solve it to a tight tolerance since its coefficients and sources are only approximate based on the values of the dependent variables from the previous iteration or time step. These iterations are called inner iterations.

The sequence is then repeated in the outer iterations by updating the coefficient matrix and the source terms until the solution converges.

$$\mathbf{r}_\phi = A_\phi \boldsymbol{\phi} - \mathbf{b}_\phi \qquad (2.70)$$

The convergence criterion is usually achieved when the residuals of (2.70) are reduced by three to four orders of magnitude.

2.5 Calculation of Screw Compressor Integral Parameters

Once a solution is obtained in the form of the velocity and pressure fields within the compressor, integral parameters which quantify the screw compressor working cycle, are calculated.

Integral parameters are used to evaluate and compare the processes in screw machines and to serve as input parameters for the design of these machines. They are divided into two groups; those based on the compressor delivery, which is the volume flow calculated at the suction conditions, and those based on the compressor power input. Other integral parameters are calculated from the previous two. These are specific power, volumetric and adiabatic efficiencies, load on the compressor rotors and bearings, torque on the male and female rotors and oil flow. Apart from these, the indicator diagram can be calculated from the pressure distribution within the compressor working cycle.

The volume flow is calculated at the inlet and at the outlet of the screw compressor by the use of the Gauss divergence theorem to calculate flow from the velocity in each particular cell in a cross section and then to integrate all of them over the complete cross section, or by integration of the mass sources along the inlet and outlet receivers.

The Gauss divergence theorem:

$$\int_s \mathbf{v} \cdot d\mathbf{s} = \int_V div \mathbf{v} \, dV \qquad (2.71)$$

This equation is integrated over a layer of cells in the cross section of the inlet or outlet port to get:

2 Computational Fluid Dynamics in Screw Machines

$$\dot{V}_f^{(t)} = \sum_{i=1}^{I} v_f S_{fi} \qquad (i = 1, 2, \ldots, I) \qquad (2.72)$$

where index f represents the direction of flow. S and v are the cell surface area and the velocity component in the direction of fluid flow. Equation (2.72) is calculated for each time step in the compressor working cycle and integrated over the complete cycle to estimate the volumetric flow at that cross section pressure and temperature:

$$\dot{V} = 60 \cdot \sum_{t=t_{start}}^{t_{end}} \dot{V}^{(t)} \quad \left[m^3/\min \right]. \qquad (2.73)$$

The volume flow obtained from this equation can be compared with the volume flow calculated from the mass added to or subtracted from the inlet and outlet receivers. These two should be the same for each time step as well as for the complete compressor cycle.

The mean density values for equal cell volumes are calculated for each time step together with the main flow:

$$\overline{\rho}^{(t)} = \frac{\sum_{i=1}^{I} \rho_i}{I} \qquad (2.74)$$

If the mean density is multiplied by the corresponding volume flow it gives the mass flow in its integral form as:

$$\dot{m} = \sum_{t=t_{start}}^{t_{end}} \dot{V}^{(t)} \cdot \overline{\rho}^{(t)} \qquad (2.75)$$

The compressor mass flow is calculated separately for the inlet and outlet chambers and these values must be identical for steady flow conditions. If they differ, then the procedure has not converged.

Another group of variables is based on the value of pressure in the working chamber.

A cell on the rotor boundary is shown in Figure 2-3. The pressure in the cell generates the force on the boundary surface. That force is calculated as the product of the pressure in the rotor boundary cell and the boundary cell surface area. This force can be also divided in three components acting in the x, y and z directions of the absolute coordinate system. When calculated, these three components are:

$$F_x = p_b A_{xb}; \qquad F_y = p_b A_{yb}; \qquad F_z = p_b A_{zb}, \qquad (2.76)$$

2.5 Calculation of Screw Compressor Integral Parameters

where p_b is the pressure in the boundary cell and A_{xb}, A_{yb} and A_{zb} are projections of the boundary cell surface in the main directions of the absolute coordinate system.

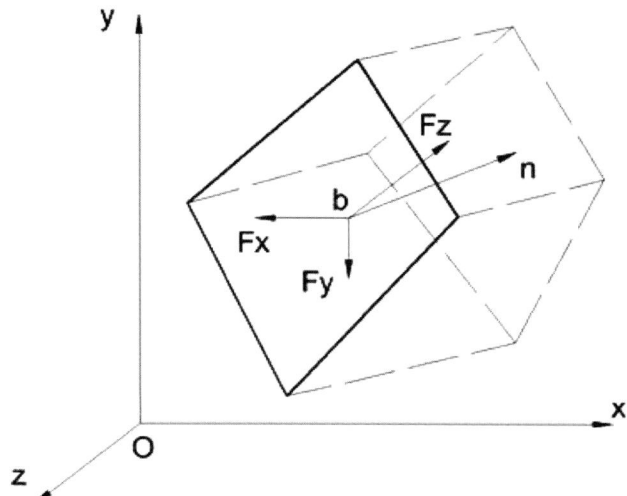

Figure 2-3 Pressure forces on the boundary surface

A free body diagram, with all pressure forces acting on a cell face and the restraint forces, is shown in Figure 2-4.

The balance for both, male and female rotor is expressed by the same set of equations:

$$\begin{aligned}
\Sigma F_x = 0: & \quad F_{xD}(i) + F_{xS}(i) - F_x(i) = 0 \\
\Sigma F_y = 0: & \quad F_{yD}(i) + F_{yS}(i) - F_y(i) = 0 \\
\Sigma F_z = 0: & \quad F_a(i) - F_z(i) = 0 \\
\Sigma M_x = 0: & \quad -F_{yD}(i)l + F_y(i)z(i) - F_z(i)y(i) = 0 \\
\Sigma M_y = 0: & \quad F_{xD}(i)l - F_x(i)z(i) + F_z(i)\left[x(i) - a\right] = 0 \\
\Sigma M_z = 0: & \quad T + F_x(i)y(i) - F_y(i)\left[x(i) - a\right] = 0
\end{aligned} \quad (2.77)$$

In these equations l is the rotor length. This set of equations applies both to the male rotor, where $a=0$ and to the female rotor where a is equal to the distance between the centre lines of the rotor axes.

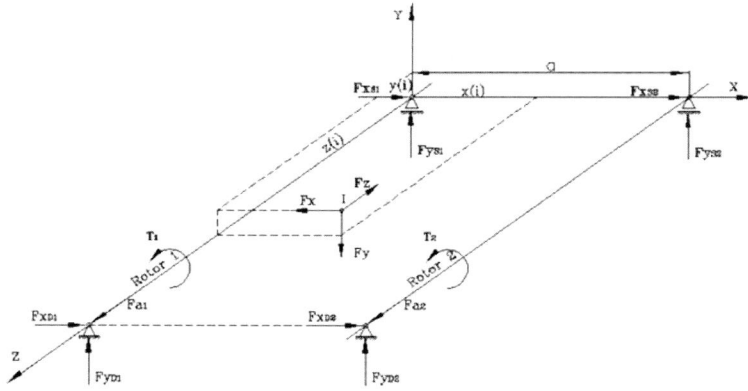

Figure 2-4 Restraint forces and torques on rotors

The torque and restraint forces on the suction and discharge bearings are calculated from these equations as:

$$F_{xS}(i) = \frac{F_y(i)z(i) - F_z(i)y(i)}{l} \qquad F_{xD}(i) = F_x(i) - F_{xS}(i)$$

$$F_{yS}(i) = \frac{F_x(i)z(i) - F_z(i)[x(i)-a]}{l} \qquad F_{yD}(i) = F_x(i) - F_{xS}(i) \qquad (2.78)$$

$$F_{rS}(i) = \sqrt{F_{xS}^2(i) + F_{yS}^2(i)} \qquad F_{rD}(i) = \sqrt{F_{xD}^2(i) + F_{yD}^2(i)}$$

$$T(i) = F_x(i)y(i) - F_y(i)[x(i)-a] \qquad F_a(i) = F_z(i)$$

All the forces in equation (2.78) are support forces caused by the pressure loads in one boundary cell i. To obtain the integral radial and axial forces and the torque they are integrated over the whole boundary and for both rotors:

$$F_{rS} = \sum_{i=1}^{I} F_{rS}(i), [N] \qquad F_{rD} = \sum_{i=1}^{I} F_{rD}(i), [N] \qquad (2.79)$$

$$F_a = \sum_{i=1}^{I} F_a(i), [N] \qquad T = \sum_{i=1}^{I} T(i), [Nm]$$

Once calculated, the torque is used to estimate the power transmitted to the rotor shaft:

$$P = 2\pi(n_m T_m + n_f T_f), \; [W] \qquad (2.80)$$

where n is the speed of the male rotor while T_M and T_F are the torque on the male and female rotors respectively. Specific power is defined as:

2.5 Calculation of Screw Compressor Integral Parameters

$$Pspec = \frac{P}{\dot{V} \cdot 1000} \quad , \left[\frac{kW}{m^3 \min}\right] \tag{2.81}$$

And finally, the values η_v and η_i, the volumetric and adiabatic efficiencies respectively are:

$$\eta_v = \frac{\dot{V}}{V_d} \qquad \eta_i = \frac{P_{ad}}{P} \tag{2.82}$$

where V_d is the theoretical displacement and P_{ad} is the adiabatic power input.

Since the pressure across the working chamber does not vary too much within one time step, it is sufficiently accurate to average the pressure values arithmetically in each working chamber in order to plot a pressure versus shaft angle, (p-α) diagram.

3

Grid generation of Screw Machine Geometry

3.1 Introduction

The finite volume method, described in the last chapter is a convenient technique which allows fast and sufficiently accurate solution of the governing equations for fluid flow within complex geometries. The resulting system of algebraic equations remains conservative within a calculation volume. To use this method a spatial domain must be replaced by a finite number of discrete volumes constructed between grid points. The number of control volumes used to discretise a computational domain depends on the geometry and physics of the problem and on the required accuracy. The number of grid points used to construct the numerical cells depends on the dimensionality of the problem and on the type of numerical cell selected for calculation. For two-dimensional cases, numerical cells can be constructed from three or more numerical points, but rectangular and triangular cells are those mostly used in practice. If the problem is three-dimensional, the most frequently used cells are hexahedral and tetrahedral volumes constructed around four and eight numerical points respectively. However, volumes may be constructed from an arbitrary number of faces for which a corresponding number of points can be calculated. The use of hexahedral cells gives the most conservative interpolation of values in neighbouring cell centres and is therefore the preferred grid type. This is especially important when geometrical and physical parameters vary substantially across a domain, as is the case in a screw machine.

The process of replacing a spatial domain by a system of grid points is called grid generation. The grid generation process, referred to as space discretisation, is essential for accuracy, efficiency and ease of numerical solution. Ability to generate an 'acceptable' grid system is a factor which determines whether a three-dimensional numerical method, such as the finite volume, can be used. Inability to generate an adequate numerical mesh for screw machines was the main reason why they have not been previously analysed by the use of three-dimensional numerical methods. The grid generation process is performed in the selected coordinate system with constraints given for a specific problem. Therefore, a coordinate system must be introduced and the required constraints have to be set. An absolute Cartesian coordinate system is used in the grid generation process because it gives the best results for the finite volume method.

3 Grid generation of Screw Machine Geometry

Grid generation problems are mainly connected with computational fluid dynamics but, as reviewed by *Thompson* (1996), the applicability of the concepts used in making numerical grids is not in any way limited to this area. Several international conferences have been held on grid generation. Conference Proceedings edited by *Smith* (1992), *Eiseman et al* (1994) and *Soni et al* (1996) are typical examples of these and describe a number of applications to CFD and other fields, especially computer graphics.

There are four major textbooks on grid generation. The early monograph by *Thompson, Warsi and Mastin* (1985) is now also published on the Web http://www.erc.msstate.edu/education/gridbook. It covers the entire subject of grid generation developed around structured grids with boundary conforming. A later book by *Knupp and Steinberg* (1993) gives fundamentals of grid generating techniques. *Thompson's* (1999) Handbook of Grid Generation describes the principles of all structured, unstructured and hybrid grids that can be generated either analytically or by solving partial differential equations (PDE). It also gives a very detailed review of most of the major computer codes used for meshing. A book on 'Grid generation Methods' by *Liseikin* (1999) was originally printed in Russian by the Siberian Branch of the Russian Academy of Sciences in Novosibirsk and then translated into English and printed in Germany and the USA. It gives a most detailed mathematical basis of grid generation methods with all necessary theoretical backgrounds. It pays special attention to reviewing the most recent and promising approaches and methods, which have not been sufficiently covered in previous monographs.

A considerable number of general methods for structured grid generation have been reviewed by *Thompson* (1984), *Thompson, Warsi and Mastin* (1985), *Eiseman* (1985), *Liseikin* (1991), *Thompson and Weatherhill* (1993) and *Thompson* (1996). The last one was also published on the Web in year 2000: http://www.erc.msstate.edu/~joe/gridconf. All these consider an algebraic approach to grid generation as well as to the solution of either elliptic, hyperbolic or parabolic partial differential equations. From these, it can be deduced that more control of numerical mesh orthogonality and smoothness is achievable by solving the PDE than by using an analytical approach. However, these methods are generally more computer time consuming then the algebraic ones.

Although, officially, transfinite interpolation for meshing purposes was first introduced at the grid conference in Nashville in 1982 by *Gordon and Thiel,* algebraic grid generation, which uses a transfinite interpolation method, has its roots in the early work by *Gordon* (1969, 1973). Many authors after that referred to their achievements in analytical methods. It started with *Smith's* Algebraic Grid Generation in 1982 when Hermite interpolation between two opposite boundaries was presented. Then *Vinkour* (1993) introduced one-dimensional stretching functions. Two, four and six boundary interpolation which use one-dimensional stretching functions were well summarised by *Shih et all* (1991) and later applied to complex grids using enhanced control of grid distribution by *Steinhorsson, Shih and Roelke* (1992). Their latest paper proposed the use of multidimensional stretching functions, applied to two opposite boundaries in the four-boundary method. Other simpler methods were also developed and reported such as that of *Saha and Basu*

(1991), but they were applicable only to rather simple grids. Orthogonality and clustering control for algebraic grid generation was reported by *Chawner and Anderson* (1991). An algebraic homotopy procedure for 2 and 3-D domains was introduced by *Moitra* (1992). *Zhu, Rodi and Schoenung* (1992) suggested a fast method for smooth grids by algebraic transfinite interpolation. Some of the methods previously reported were revised in the paper given by *Soni* (1992) and applied to internal flow configurations. More recently, *Liou and Yeng* (1995) presented an algebraic method which generates a numerical grid by combining two grids and uses a marching method in calculating the final mesh. Zhou (1998) presented a 'simple' method using functions of the first, second or third order to interpolate between the boundaries. The method, although simple and fast, often produces overlapping and irregularity for complicated domains. The main problem that remained during all these years of development of algebraic methods was that meshes generally have not been orthogonal and smooth and that overlapping appeared as the major problem in complex domains. More recently, *Lehtimaki* (2000) suggested another method in which the first step is to generate a correct mesh without overlapping, not necessarily smooth and orthogonal, and then to apply orthogonalization and smoothing to it by moving internal points. The method appeared to be extremely useful but required additional checking of the mesh regularity. *Field* (2000) in his journal paper presented qualitative measures for initial meshes, which, together with a good description and mathematical background of the grid quality measures by *Liseikin* (1999), reasonably covers the problem.

Differential methods for the structured grids to solve elliptic, parabolic or hyperbolic equations were discussed by many authors and were well summarised by *Thompson* (1985, 1996 and 1999) and *Liseikin* (1999).

Surveys on general adaptive methods were presented by *Eiseman* (1985), *Liseikin* (1996, 1998 and 1999) and *Thompson* (1999). Practical applications of adaptive methods have been published by *Kim and Thompson* (1990) and *Samareh-Abolashi and Smith* (1992). The latest described a practical approach to the grid adaptation based on two physical properties. The method was used for dynamic grid adaptation in supersonic flow domains.

The methods for unstructured grids were revised by a number of authors among whom are *Thompson* (1985, 1999), *Liseikin* (1999), *Eiseman* (1985) and *Owen* (1998). The latter published his review on unstructured grid methods on the Web (http://www.andrew.cmu.edu/user/sowen/survey/index.html) where a summary on recently developed methods in automatic generation for complicated domains is given. Among these methods, the major ones are conformal mapping by *Barker and Lantz* (1997), control point grid generation by *Eiseman* (1991 and 1992) and quad-mapping by *Owen et al* (1998).

More recently a huge effort has been made to develop grid generation in Cartesian co-ordinates by *Lin and Chen* (1998). Its main advantage lies in the fact that the conservation laws are completely satisfied, without the need for additional terms, but they have a great disadvantage in the need for mesh refinement in the vicinity of the boundaries.

3.1.1 Types of Grid Systems

Grid systems are structured, unstructured or mixed, depending on how the grid points are connected to each other. These types are presented in Figure 3-1. Unstructured numerical grids are easier to generate. However, obtaining a solution with them is more difficult then from structured grids.

Structured grids are either single or composite grids. A single grid is shown in Figure 3-2. A composite grid consists of two or more single grids patched together. Depending on how the individual single grids are assembled together, a composite grid system can further be classified as completely discontinuous, partially discontinuous, partially continuous and completely continuous as shown in Figure 3-2.

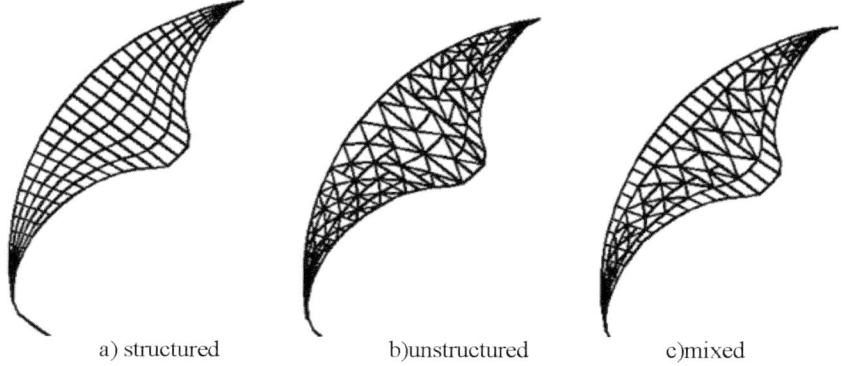

a) structured   b) unstructured   c) mixed

Figure 3-1 Types of a grid system

Figure 3-2a shows an overlapping or Chimera grid. This grid is easy to construct and implement for complex geometries. However conservation is not always achieved on block boundaries. The patching process requires only the generation of a single grid over each compressor rotor without considering the interface between them. The constraint in such a case is the amount of overlap between the grids on the rotors. Another advantage of this grid is that it allows separate computation on each grid, which preserves the computer resources. The disadvantage of this procedure is that conserved monotonic values have to be maintained in the grid points of the overlapping regions, during the solution and this is not always possible.

Partially discontinuous composite grids are usually called block-structured grids with a non-matching interface. Grids of this type are generated by partitioning the domain of interest into a number of non-overlapping continuous blocks. Then a structured numerical mesh is generated within each of the blocks. After that the blocks are patched together through the non-overlapping interface. This grid is flexible and it allows generation of the grid with the desired resolution and topology in each particular block which can be later refined in a block wise manner. Moreover, methods for conservative treatment of a non-matching interface already exist and these are already implemented in standard CFD software. Block structured grids are used for spatial discretisation of screw machine rotors.

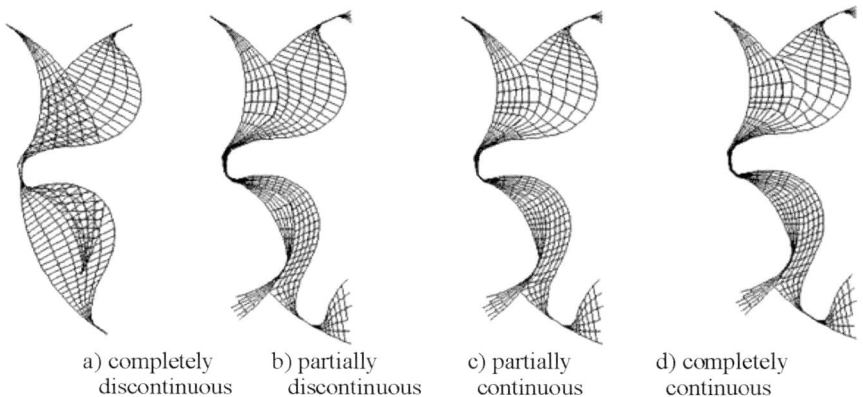

a) completely discontinuous   b) partially discontinuous   c) partially continuous   d) completely continuous

Figure 3-2 Types of composite structured grids

Partially and fully continuous composite grids are difficult to generate. Their points at the matching interface coincide. In the case of a three-dimensional domain this is difficult to achieve. These grids can numerically be treated as a single grid because there is no real interface between them. The conservation of properties is as well satisfied as for a single grid. These types of grid are used for the suction and discharge chambers and for other sub domains of a screw machine other then the rotors. A structured grid system, either single or composite must satisfy a number of conditions.

The total number of grid points should be kept to the minimum needed for the numerical method to yield a solution of the desired accuracy. The numerical points which form the boundary grid lines should coincide with the boundary of the spatial domains. If possible, one set of grid lines should align with the flow direction. This condition is not easy to fulfil, but it helps to obtain an accurate numerical solution. The grid lines should be perpendicular to the boundary. This allows easier implementation and calculation of derivatives at the boundaries of the domain. In the interior, the intersecting lines need be only nearly orthogonal. The spacing between the grid points should change slowly from a region with concentrated points to a region with sparsely distributed nodes, especially if the gradients of the flow values are large in these regions.

The most efficient structured grids are boundary-fitted or boundary-confirming grids. These grids are formed in such a manner that one set of points is fitted to the boundary of a domain so that the boundary conditions are directly applied to the boundary region without the need for interpolation. In that case, the boundary conditions are considered as input data.

It is common practice to generate and distribute numerical points on physical boundaries and then extend them successively from the boundary to the interior of the domain. Based on this, three basic groups of grid generation methods have been developed: algebraic methods, which use interpolation or some special functions, differential methods based on the solution of partial differential equations in the transformed spatial region, and variational methods, based on optimisation of the grid quality properties.

Algebraic methods calculate interior points of the grid through transfinite interpolation, which is a multivariate interpolation procedure. They are relatively simple and they enable the grid to be generated quickly. However, in regions with a complicated shape, cell faces generated by algebraic methods can degenerate so that cells may overlap or cross the boundary. These methods are therefore commonly used to generate grids in domains with smooth boundaries which are not highly deformed, or as an initial approximation for iterative processes in a differential grid solver. Algebraic methods generate a screw compressor numerical grid of the desired quality, if used in conjunction with boundary adaptation and procedures to obtain orthogonal grids.

Differential methods are iterative procedures to solve elliptic, parabolic or hyperbolic partial differential equations, of spatial point distribution. The interior coordinate lines derived through these methods are smooth, assuring that discontinuities from the boundary surface do not extend into the domain interior. In practice, hyperbolic systems are simpler then elliptic or parabolic ones but these are not always mathematically correct and are not applicable to regions in which a complete boundary surface is strictly defined.

Variational methods can be used to generate grids which satisfy more then one generating condition, that cannot be satisfied either by algebraic or differential methods. However, these are not widely used mainly because their formulation does not always lead to a well posed mathematical problem.

3.1.2 Properties of a Computational Grid

A numerical grid should divide a physical domain to enable efficient computation of the physical quantities. The accuracy is influenced by the grid size and its expansion factor. The grid size is determined by the number of grid points, while the cell size implies the maximum length of the cell edges. A grid generation procedure should produce a mesh for an arbitrary number of cell nodes or for an arbitrary cell size in such a way that the cell size is reduced if the desired number of nodes increases. This property is important for obtaining an accurate solution which might require refined cells to be generated in specific domains of interest. The ability to increase the number of grid points and to reduce the size of a cell enables the convergence rate to be increased and the accuracy enhanced. The important factor is a grid cell shape, quantified through the expansion factor f_e, which is the ratio of dimensions in two neighbouring cells. It is difficult to suggest a maximum value for this factor, because it depends on local grid topology and oscillation of the dependent variables in the considered region.

Grid orthogonalisation and smoothness are obtained by algebraic generation to achieve a better numerical solution. The finite volume method is not so sensitive to grid non-orthogonality, but some orthogonality and smoothness are recommended, especially at boundaries. It allows easier implementation of boundary conditions and increases the stability of the calculation, especially at the interfaces between grid blocks.

Cell and grid deformation is a measure of departure from a standard and non-deformed cell. It is defined through at least three values, the aspect ratio, angle of non-orthogonality and a warp angle, as shown in Figure 3-3. Both the numerical accuracy and the stability are dependent on these three factors. The aspect ratio f_a is between the longest and the shortest edge of the numerical cell and it should be close to unity. The non-orthogonality angle θ_{no} is between the surface vector and the distance vector, which connects the centres of two neighbouring cells that share the surface. Its value should be close to zero. The warp angle θ_w measures non-coplanarity of the cell face. For coplanar cell faces, the vectors are parallel and the warp angle is zero. Otherwise, the angle between the surface normal to the two triangular subsurfaces differs from zero.

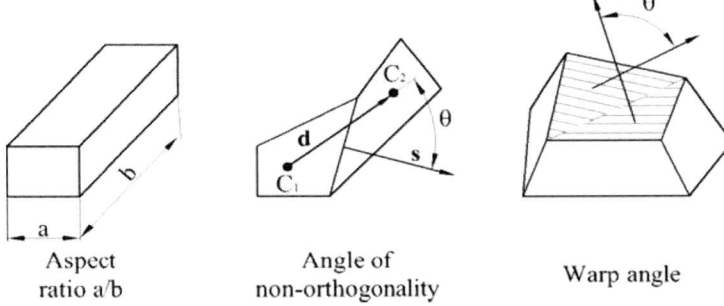

Aspect ratio a/b   Angle of non-orthogonality   Warp angle

Figure 3-3 Measures of grid quality

Although the limiting values for mesh quality measures are different for each particular problem, the following values are recommended:

    Expansion factor $fe < 2$
    Aspect ratio $f_a > 0.1$
    Angle of non-orthogonality $\theta_{no} < 50°$
    Wrap angle $\theta_w < 50°$

Consistency with geometry is a mesh property that significantly influences the result. It has two important aspects. The first is that the numerical mesh must have a sufficient number of points in the interior of the computational domain, which accurately describe the physical domain. When the number of points increases to infinity, the point distance should reduce to zero. This is important for a screw machine where the domain size changes by several orders of magnitude between the clearance and the main domain while the number of points is kept constant. The second requirement is that a sufficient number of points is specified on the boundary. This can be achieved by increasing the number of points or by adjustment of the boundary according to its geometry. With both requirements fulfilled, the numerical mesh is boundary fitting or boundary confirming.

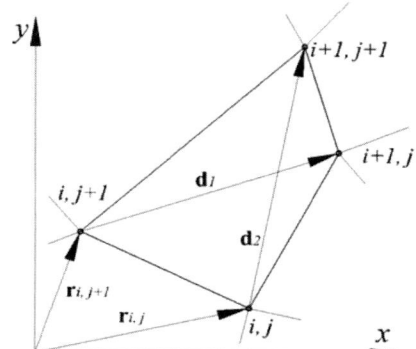

Figure 3-4 Vector definition of a two-dimensional cell

If the grid is generated by an analytical procedure before carrying out the flow calculations, then it cannot be dynamically adapted during the calculation of the physical values associated with the flow. Hence it should be set initially with finer resolution in the regions of high flow gradients in order to maintain consistency between grid generation methods and the physics of the flow. This condition is difficult to fulfil at the beginning of the grid generation or prior to obtaining a solution. Therefore, based on the available data, it must be possible to adjust the generated grid to both the geometry and the flow regime. Generally, if the lines connecting numerical points follow the direction of the flow and if the distance between the points is smaller in the regions near the walls, this condition may be fulfilled.

Finite volume methods, in the current state are capable of producing a solution with both, structured and unstructured grids. However, conservation principles are easier to obtain on structured meshes with hexahedral cells. Therefore, these should be generated wherever possible. This gives the possibility of connecting the generated numerical mesh to a wider range of numerical solvers.

The quality of different meshes may be assessed by different criteria. One of these is the skewness value, which is calculated to give a measure of the quality of each particular cell and of the mesh as a whole. If a two-dimensional cell is considered, the skewness value σ_s is defined as the cell area divided by the maximum lengths of cell edges in two curvilinear coordinate directions as shown in Figure 3-4:

$$\sigma_s = \frac{A}{e_1 e_2} \qquad (3.1)$$

where

$$e_1 = \max\left\{\left|\mathbf{r}_{i+1,j} - \mathbf{r}_{i,j}\right|, \left|\mathbf{r}_{i+1,j+1} - \mathbf{r}_{i,j+1}\right|\right\}; \quad e_2 = \max\left\{\left|\mathbf{r}_{i,j+1} - \mathbf{r}_{i,j}\right|, \left|\mathbf{r}_{i+1,j+1} - \mathbf{r}_{i+1,j}\right|\right\}$$

and the cell area is calculated as half the vector product of the diagonals:

$$A = \frac{1}{2}\left|\mathbf{d}_1 \times \mathbf{d}_2\right|.$$

The sign of the skewness factor indicates whether a numerical cell is regular or not. Positive values are obtained for regular cells, while negative skewness indicates that the cell is inverted or twisted. At the same time, this indicates whether the cell is orthogonal or not. For an orthogonal cell the skewness value $\sigma_s=1$. Otherwise, the value of the skewness factor is less then 1 and tends to zero for extreme non-orthogonality. For three-dimensional cells, the skewness can be calculated as the ratio of cell volume to the values of the maximum edge lengths in the three curvilinear coordinate directions.

Grid quality indicators are calculated for every cell in a computational domain. The mesh can be accepted as regular only if all cells have a skewness value greater then zero.

3.1.3 Grid Topology

Block structured grids are convenient for the grid generation of complex geometries. But although the grid generation process is simplified when the whole domain is subdivided into a number of simpler blocks, it is not always easy to select a suitable grid topology within a block. Also, although simpler for a whole domain, a sub-domain is not necessarily an efficient means of grid generation. The aim of algebraic grid generation is to find a function $x(\xi)$ which transforms a computational domain Ξ^n to the physical domain $\mathbf{X}^n$ or vice versa, as shown in Figure 3-5. Four basic topologies are used to specify a numerical grid within the block. These are polyhedral, H, O and C grids. Only polyhedral and O grids have been used in screw machine grid generation.

a) A block type grid is represented as a polyhedron, which retains the schematic form of a block domain. This type is mainly used for single-block grids. A numerical grid has all the properties of a computational block and fulfils the requirements of a physical domain by boundary fitting to the computational domain. It is used here to produce grids of the inlet and outlet ports as well as other regions which retain a polyhedral block shape. Both physical and numerical domains keep their block shape as shown in Figure 3-5a. The numerical domain is transformed in a hexahedral block while the physical domain remains fitted within the specified boundaries.

48   3 Grid generation of Screw Machine Geometry

b) An O type of numerical grid is generated as a solid cube or rectangle in the computational domain and transformed to the physical domain as presented in Figure 3-5b. This type of numerical grid has only two boundary faces in two dimensions and four boundaries in 3D. The remaining two boundary faces are connected implicitly point to point.

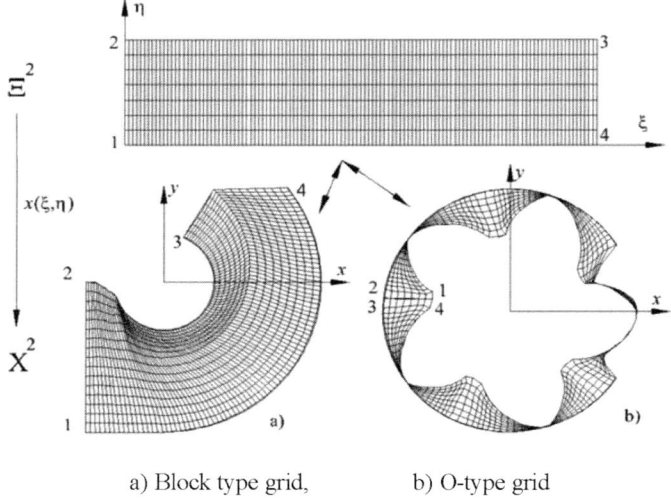

a) Block type grid,   b) O-type grid

Figure 3-5 Patterns of grid topology in a screw machine

H and C grid types are constructed to a block grid by a similar procedure, with the required shape formed by explicit connectivity between required points.

3.2 Decomposition of a Screw Machine Working Domain

To apply a CFD procedure, the spatial domain of a screw machine is replaced by a grid with discrete finite volumes and a composite grid, made of several structured grid blocks patched together and based on a single boundary fitted co-ordinate system, as shown in Figure 3-6. The number of these volumes depends on the problem dimensionality and accuracy required. It consists of several sub-domains, two of which are critical. These are the fluid domains around the male and female compressor rotors. Not only is the main working domain contained in them but also all the clearances and leakage paths, such as the radial, axial and interlobe leakage gaps and the blow-hole area. The suction and discharge and other ports, such is the oil injection port, are each presented by an additional block. The grid blocks are then connected over the regions on their boundaries which coincide with the other parts of the numerical mesh.

3.2 Decomposition of a Screw Machine Working Domain 49

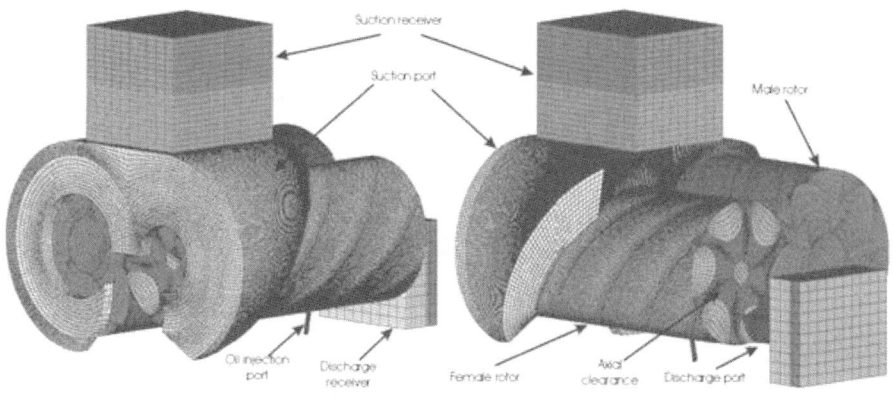

a) Suction side   b) Discharge side
Figure 3-6 Numerical mesh of a screw compressor

The rotors of a screw machine are helical surface elements generated by simultaneous rotation and translation along the rotor axis, that are based on a two-dimensional definition of the profile points in cross section. The rotors are the main machine parts between which the compression process takes place. Therefore, these are studied in more detail here, while other parts of the numerical mesh are explained only briefly.

Grid generation of screw machine rotors starts from the rotor profile coordinates and their derivatives. These are obtained by means of a generation procedure. Interlobe clearances are accounted for by the geometry and added to the rotors. They depend on the application of the screw machine and their distribution is specified in advance.

The envelope meshing method, *Stosic* (1998) is applied for generation of both the male and female rotor profiles as shown in Figure 3-7a. The rotor geometry is completed by generating the profile around the rotor axis for an angle defined by the number of lobes:

$$\varphi_{r1} = i \cdot 2 \cdot \pi / z_1 ; \qquad \varphi_{r2} = -\varphi_{r1} \cdot z_1 / z_2 , \qquad (3.2)$$

where z_1 and z_2 are number of lobes on the male and female rotors respectively. By this means, the rotor geometry is obtained for each cross section, together with the inner boundary of the "O" mesh on the rotors.

To obtain the outer boundary of the "O" mesh, the rotor rack is used and connected to the outer rotor circle to form a closed line as shown in Figure 3-7b. The circle, to which the rack is connected, represents the rotor housing. It is formed by adding a radial clearance to the outer rotor circle r_o:

$$r_{1_o} = r_{1_e} + \delta_{r_1} ; \qquad r_{2_o} = r_{2_o} + \delta_{r_2} . \qquad (3.3)$$

50   3 Grid generation of Screw Machine Geometry

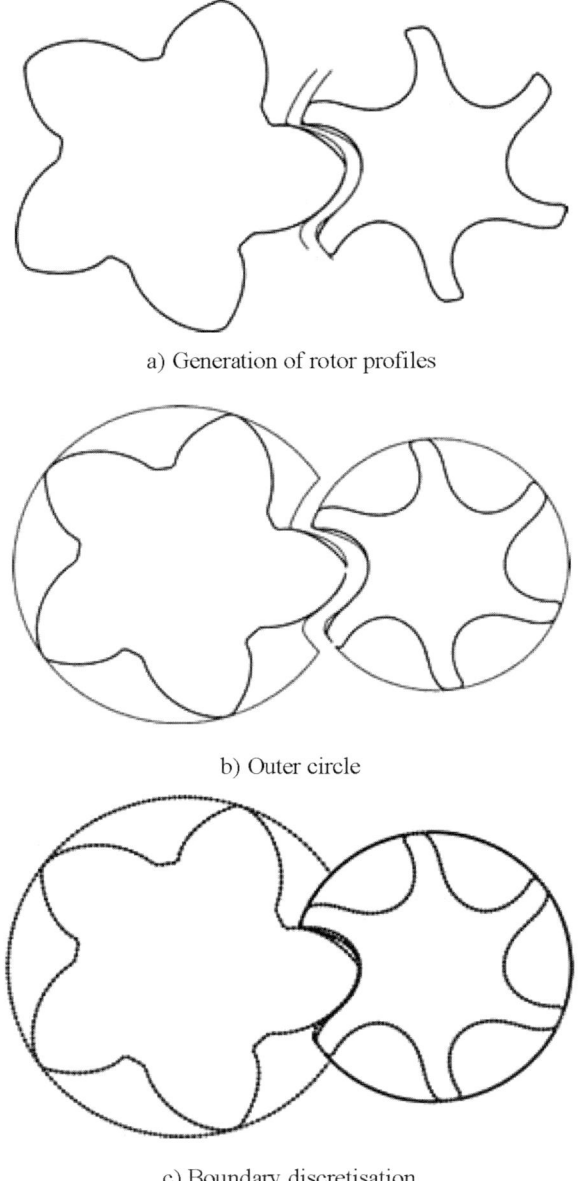

a) Generation of rotor profiles

b) Outer circle

c) Boundary discretisation

Figure 3-7 Phases in the generation of screw compressor rotor boundaries

The boundary points then may be set at a constant distance along the profile. However, in such a case, some details of the rotor profile may be lost. Therefore, the boundaries are discretised according to the rotor geometry particulars and flow characteristics. The distribution is applied to the inner boundary to follow the rotor coordinate points. The regular point distribution on the outer boundary assures a proper generation of inner grid points.

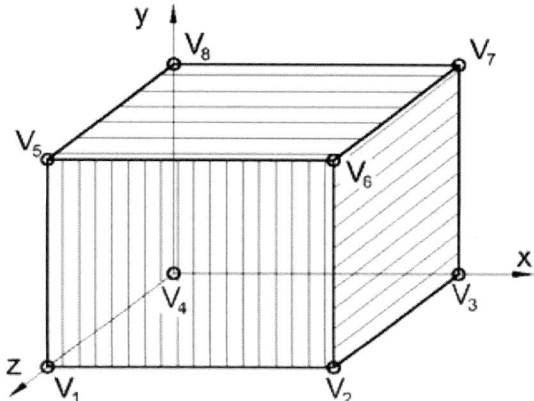

Figure 3-8 Hexahedral numerical cell

When both boundaries are mapped by an equal number of boundary nodes, an analytical transfinite interpolation is applied to generate the internal points of the numerical mesh. The screw machine rotor geometry is described by the boundary and internal points in a sequence of rotor cross sections. A 3-D numerical mesh is then formed by connecting these points in consecutive cross sections. Preferably, the entire numerical mesh is formed of hexahedral control volumes with a right-hand definition in which eight grid points form a right-handed coordinate system as in Figure 3-8. However, if hexahedral cells are not attainable, the numerical cells may degenerate to form either a prism, a pyramid or a tetrahedron, as shown in Figure 3-9. In the case of the degenerated cells, when the edges collapse, the co-inciding points are retained to keep their index.

After the cells of the entire numerical mesh are defined, the boundary regions are defined through which the domains are connected or at which the boundary conditions are applied.

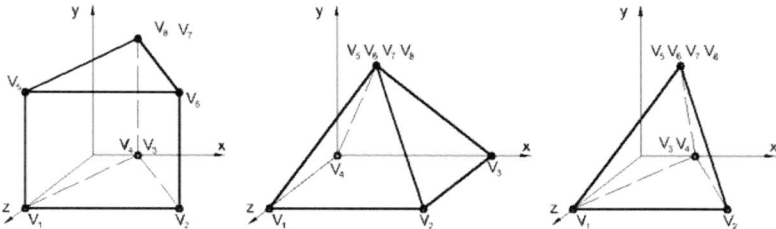

Figure 3-9 Degenerated cell shapes: prism, pyramid and tetrahedron

Region elements of either a quadrilateral or triangular shape are formed, depending on the shape of the boundary cell face, by using the vertices of the boundary cell face. This defines the numerical mesh by the vertex, cell and region specifications. This form of data allows convenient connection to a general numerical solver of the finite volume type.

3.3 Generation and Adaptation of Domain Boundaries

Grid adaptation is achieved by the use of differential or algebraic methods. Differential methods are based on the simultaneous solution of the Euler-Lagrange differential equations and the fluid flow equations when the mesh is adapted. Algebraic methods are based on a direct equidistribution technique which does not require solution of the differential equations.

Numerical grid adaptation can be performed dynamically or statically. Dynamic adaptation is mainly used for the simulation of fluid flow with high local gradients. It is applied together with the calculation of the flow properties. Static grid adaptation is usually used to improve a numerical grid in advance, before obtaining a flow solution. Such adaptation is based on the existing boundary geometry and on the flow characteristics expected in the domain.

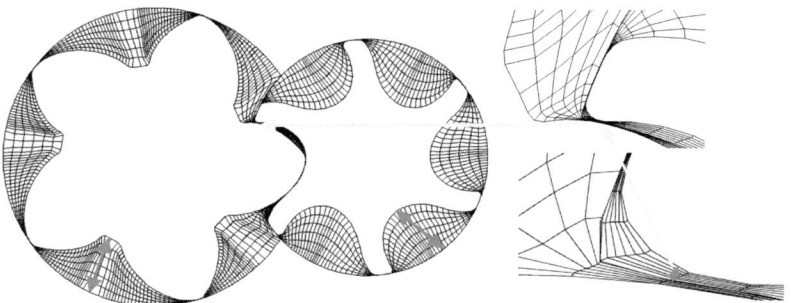

Figure 3-10 Comparison of cell sizes in the working chamber and clearances

A static analytical adaptation is applied here to the rotor boundaries of a screw machine. Firstly, this is because the cells in the clearance regions are excessively deformed as a consequence of their being equal in number in the radial direction to those in the working cavity, as shown in Figure 3-10. As a result their aspect ratio can be as high as 1000:1. Therefore, the cell shapes easily become too deformed for accurate calculation. Secondly, if a low number of boundary points is applied to a very curved boundary, for example on the top of the female rotor, some of the geometry features which significantly affect flow, may be lost. This situation can be overcome by introducing more cells along the boundary or alternatively by local mesh refinement. Both approaches lead to an increase in the number of control volumes, which adversely affects both the efficiency and speed of calculation.

The approach adopted here distributes numerical points selectively with more, for example, in the clearances, and less in other places. The numerical mesh in Figure 3-10 is generated from adapted 2-D rotor boundaries where the ratio of the working chamber to clearance size was 400:1.

3.3.1 Adaptation Function

The analytical equi-distribution technique minimises the error by redistributing points along a curve to keep the product of the 'weight function' and the grid spacing constant, i.e.

$$X_\xi \cdot W = const. \tag{3.4}$$

where X_ξ represents the grid spacing and W is the weight function.

This is the Euler-Lagrange equation. When integrated with the respect to a computational coordinate ξ it becomes:

$$X(\xi) = \int_0^\xi \frac{d\xi}{w(\xi)} \bigg/ \int_0^1 \frac{d\xi}{w(\xi)}. \tag{3.5}$$

In equation (3.5), both the grid spacing and the weight function depend on the computational coordinate ξ. This means that the equation is implicit and must be solved iteratively.

However, if equation (3.4) is integrated with respect to the physical coordinate x, it has an explicit form which can be solved directly. According to *Samareh-Abolhassani and Smith* (1992) this is:

$$\xi(x) = \int_{X_{min}}^{x} W(x)dx \bigg/ \int_{X_{min}}^{X_{max}} W(x)dx. \tag{3.6}$$

Finally, if the starting equation is integrated with respect to the natural coordinate or an arc length, which follows a curve, its form is:

$$\xi(s) = \int_0^s W(s)ds \bigg/ \int_0^{S\,max} W(s)ds. \tag{3.7}$$

The last form of the equation is the most appropriate for adaptation of complex curves which represent the boundaries of screw machines. Its one-dimensional nature allows adaptation along one single grid line only, or along the set of grid lines which all follow a single natural direction. Multidimensional adaptation can be achieved by successive repetition of this procedure along all the sets of grid lines.

The way a grid line is adapted depends only on the selection of the weight function.

$$W(s) = 1 + \sum_{i=1}^{I} b^i f^i(s), \qquad (3.8)$$

where i is the number of variables which influence the adaptation, b^i are constants and $f(s)$ are adaptation functions or their first derivatives. The adaptation function, which appears in the same equation, is integrated along the length of the grid as:

$$F^i(s) = \int_0^S f^i(s)ds. \qquad (3.9)$$

Equations (3.7), (3.8) and (3.9) together give:

$$\xi(s) = \frac{s + \sum_{i=1}^{I} b^i F^i(s)}{S_{max} + \sum_{i=1}^{I} b^i F^i(S_{max})}. \qquad (3.10)$$

Equation (3.10) can be used to move the grid points along a fixed curve. Parameters b^i and F^i have positive values in order to be monotonic in $\xi(s)$. However, if the line, which is adapted, changes in time, or if the adaptation has to be applied to a group of grid lines in a two or three dimensional grid, b^i has to be updated to keep the same emphasis on the point concentration. In this more general case it is useful to define a grid point ratio R^j, assigned to each particular function f^j as:

$$R^j = b^j F^j(S_{max}) \Big/ \left\{ S_{max} + \sum_{i=1}^{I} b^i F^i(S_{max}) \right\}, \quad j = 1, 2, \ldots, I \qquad (3.11)$$

Now, if the grid point ratio (3.11) is calculated, continuous updating of b^i is avoided and the emphasis on the point concentration is kept constant. After implementing (3.11) to equation (3.10) the final form of the adaptation function becomes:

$$\xi(s) = \frac{s}{S_{max}} \left\{ 1 - \sum_{i=1}^{I} R^i \right\} + \sum_{i=1}^{I} \left\{ R^i \frac{F^i(s)}{F^i(S_{max})} \right\} \qquad (3.12)$$

3.3.2 Adaptation Variables

In order to map boundaries of a computational domain using the previous equation, adaptation variables must be selected, which form a weight function. Six adaptation variables with different effects are presented here but the simultaneous combination of any two of them is sufficient. The tangent angle on the curve at the calculation point, radius of curvature in the vicinity of the calculation point and distance of the calculation point from the rotor centre respectively are:

$$\alpha_P = \arctan\left(\frac{\partial x_P}{\partial y_P}\right), \quad r_P = \frac{1+y_P'^2}{y_P''} \cdot \frac{3}{2}, \quad d_P = \sqrt{x_P^2 + y_P^2}, \qquad (3.13)$$

The curve flatness around the calculation point, sinusoidal distribution of points and cosinusoidal distribution of points are:

$$E_P = \frac{\sqrt{(x_P - x_{P_o})^2 + (y_P - y_{P_o})^2}}{d_P - d_{P_o}}, \quad C_{\sin} = \sin(\pi \cdot k_P/K), \quad C_{\cos} = |\cos(\pi \cdot k_P/K)|.$$

All the above variables are geometrical characteristics, whose gradients along the boundary curve give various distributions in the characteristic region. For example, if the radius of curvature changes rapidly with the natural coordinate, then the adoption of r_P as the adaptation variable will give more points in this region than in other places. A similar situation happens if other adaptation variables are applied.

Although the computational method applied here allows adaptation by two adaptation variables, it is still general, which means that any other number of adaptation variables can be introduced for adaptation if necessary.

3.3.3 Adaptation Based on Two Variables

Any geometrical or flow characteristic can be used for boundary adaptation. For illustration of the method, the gradients of a tangent angle and radius of curvature have been used as weight functions f_k^1 and f_k^2 respectively in the first and second adaptation criterion:

$$f_k^1 = \left|\frac{\partial \alpha_P}{\partial s}\right|_k = \left|\frac{\alpha_{Pk+1} - \alpha_{Pk-1}}{s_{k+1} - s_{k-1}}\right|, \quad f_k^2 = \left|\frac{\partial r_P}{\partial s}\right|_k = \left|\frac{r_{Pk+1} - r_{Pk-1}}{s_{k+1} - s_{k-1}}\right| \qquad (3.14)$$

In the previous equation, index k counts the numerical points along the curve which is adapted. Here, weight functions are given both differentially and in the form of finite differences. Derivatives with respect to the arc-length, i.e. the physical coordinate, are the most convenient for static adaptation. However, in the case of general dynamic adaptation, derivatives with respect to the computational coordinate ξ are probably more suitable. The usual way to present them is as a central differencing on the computational mesh with equidistant distribution of the points:

$$f_k^1 = \left|\frac{\partial \alpha_P}{\partial \xi}\right|_k = \left|\frac{\alpha_{Pk+1} - \alpha_{Pk-1}}{2}\right|, \qquad f_k^2 = \left|\frac{\partial r_P}{\partial \xi}\right|_k = \left|\frac{r_{Pk+1} - r_{Pk-1}}{2}\right| \qquad (3.15)$$

The next step in modifying the boundaries is to evaluate the integral adaptation variable (3.9) which is later used to calculate the point distribution from equation (3.10). A trapezoidal rule for non-uniform spacing of the physical coordinate s_k is used here for calculation of the integrated adaptation variable at each numerical point. The discretised formula has a standard form for all applied weight functions:

$$F^i(S_1) = 0, \qquad F^i(s_k) = F^i(s_{k-1}) + \frac{\left[f_k^i + f_{k-1}^i\right] \cdot \left[s_k - s_{k-1}\right]}{2}, \qquad (3.16)$$

where $k=2,3,...,K$ is the number of points along the curve, while $i=1,2,...,I$ is the number of variables used for adaptation.

The third step is the calculation of a new computational coordinate along the curve as a function of the previous physical coordinate. This step is performed through equation (3.12), which requires the grid ratio from equation (3.11) to be known. As a result, a new coordinate distribution is given as function of the previous distribution as:

$$\xi(k) = \xi(S_k). \qquad (3.17)$$

In a more convenient form, this equation, which gives a new distribution of points, is written as:

$$\xi(k) = \frac{S_k}{S_K} \cdot \left(1 - \sum_i R^i\right) + \sum_i \frac{F^i(S_k)}{F^i(S_K)} \cdot R^i, \qquad (3.18)$$

where R^i is the weight coefficient for each adaptation variable when applying a new distribution. The sum of factors R for all adaptation variables i should be less then or equal to 1.

Finally, the fourth step is to find the inverse function of (3.17) which gives a new physical coordinate $\overline{S}_k$ in the new coordinate system $\overline{\xi}_k$. This is done by an interpolation procedure which aims to find a new arc-length in the form of

$$\overline{S}_k(\overline{\xi}_k) = \sum_{m=1}^{K} L_m(\overline{\xi}_k) \cdot S_k(\xi_k), \qquad (3.19)$$

where the overscored values represent new values of the physical and computational coordinates and the others are previous values. The new transformed coordinate and both the starting and ending values n_{min} and n_{max} of the Lagrangian product of transformed coordinates:

$$\overline{\xi}_k = (k-1)/(K-1)$$
$$L_m(\overline{\xi}) = \prod_{\substack{n=n_{min} \\ n \neq m}}^{n=n_{max}} (\overline{\xi} - \xi_m) / (\xi_n - \xi_m) \qquad (3.20)$$

must satisfy the conditions:

$$\xi_{n_{min}-1} \leq \overline{\xi}_k \leq \xi_{n_{min}}; \qquad 1 \leq n_{min} \text{ and } n_{max} \leq K. \qquad (3.21)$$

An example of the use of a geometrical weight function to adapt the rotor boundary of a screw machine is shown in Figure 3-11. In the figure on the left hand side, a pair of rotors with 4 male lobes and 6 female lobes (4/6) is shown with uniform distribution. The number of cells generated in both rotors is the same.

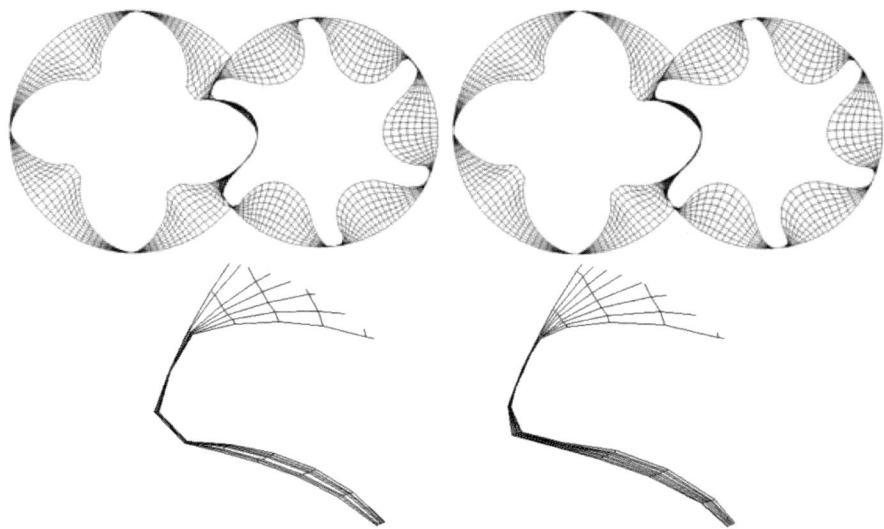

Figure 3-11 Comparison between the original (left) and adapted rotors (right)

The boundary distribution on the rotors on the right hand side is modified by two weighting functions, namely the tangent angle and the radius of curvature which values are 0.2. As a result, a better arrangement of both the boundary points and the entire mesh is obtained. This is particularly noticeable in the gaps and rotor inter-connections. However, in the parts of the numerical mesh where the cells are not so deformed and where the pattern of flow variables is nearly uniform, the cell size is not modified at all.

The distribution of boundary points plays a significant role in the grid generation process. A more appropriate boundary distribution allows easier generation of inner mesh nodes.

3.3.4 Mapping the Outer Boundary

The desired distribution on the rotor inner boundary of a 2-D "O" mesh is obtained by adapted mapping. In Figure 3-12, points are given in index notation with respect to the physical coordinate system:

$$\mathbf{r}_{i,j=0} = \mathbf{r}_{i,j=0}(x,y) \qquad (3.22)$$

The outer boundary needs be mapped with the same number of points as the inner boundary. The easiest way to obtain the same number of points on both boundaries is to apply the same arc-length distribution of points to the outer boundary as to the inner one. This successfully distributes the boundary nodes for generation of the inner points only if the mesh is simple and its aspect ratio does not change rapidly. However, for complex meshes and high aspect ratios, this method usually does not produce a distribution which results in a regular mesh. Therefore, a new arc-length on the outer boundary has to be specified before the generation of the inner points can be performed. The procedure is to transform the outer boundary to a straight line, which is then adapted to a new coordinate system.

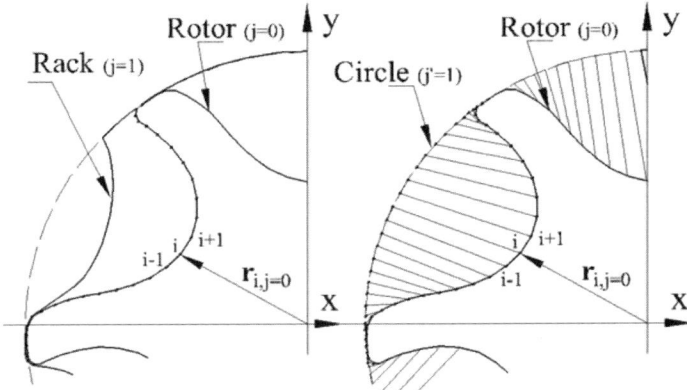

Figure 3-12 Point distribution for rotor and circle with equal arc-length

3.3 Generation and Adaptation of Domain Boundaries

This is followed by reverse transformation to the physical boundary in parametric form. To apply the procedure, the outer circle is first mapped with the same number of points and the same arc-length distribution as the rotor itself:

$$\mathbf{r}_{i,j'=1} = \mathbf{r}_{i,j'=1}(x,y) = \mathbf{r}_{i,j'=1}\left(\frac{s_i}{s_I}\right) \qquad (3.23)$$

where s_i is the natural coordinate of the rotor i.e. this is the distance from the starting point to the point i along the rotor curve, while s_I is the length of the rotor boundary.

Figure 3-12 shows how the initial distribution of the points on the outer circle is achieved by the same arc-length as on the screw compressor female rotor. The distribution of points obtained by this means is not always satisfactory especially if the points are extrapolated to a rack.

Lengths a, b and c are shown for an arbitrary point i in the physical domain, as shown on the left of Figure 3-13. These are calculated as:

$$a = \sqrt{(x_i - x_{i-1})_{j=0}^2 + (y_i - y_{i-1})_{j=0}^2},$$

$$b = \sqrt{(x_{i-1,j=0} - x_{i-1,j'=1})^2 + (y_{i-1,j=0} - y_{i-1,j'=1})^2}. \qquad (3.24)$$

$$c = \sqrt{(x_{i,j=0} - x_{i-1,j'=1})^2 + (y_{i,j=0} - y_{i-1,j'=1})^2}$$

The angle between a and c is determined by the cosine theorem:

$$\cos\alpha_{ac} = \frac{a^2 + b^2 - c^2}{2ab} \qquad (3.25)$$

The computational coordinate system ξ-η with transformed coordinates of the rotor and circle is shown on the right of Figure 3-13. The outer circle in the physical domain is transformed to a straight line along the ξ axis in the computational domain. The rotor profile is transformed from the physical to the computational domain so that each computational point on the rotor has the same value on the ξ coordinate as its corresponding point on the circle.

The point coordinates in the computational domain are given by the following expressions:

$$\begin{aligned} \xi_{i,j=0} &= \xi_{i-1,j=0} + c\cos\alpha_{ac}, & \xi_{i,j'=1} &= \xi_{i,j=0} \\ \eta_{i,j=0} &= c\sin\alpha_{ac}, & \eta_{i,j'=1} &= 0 \end{aligned} \qquad (3.26)$$

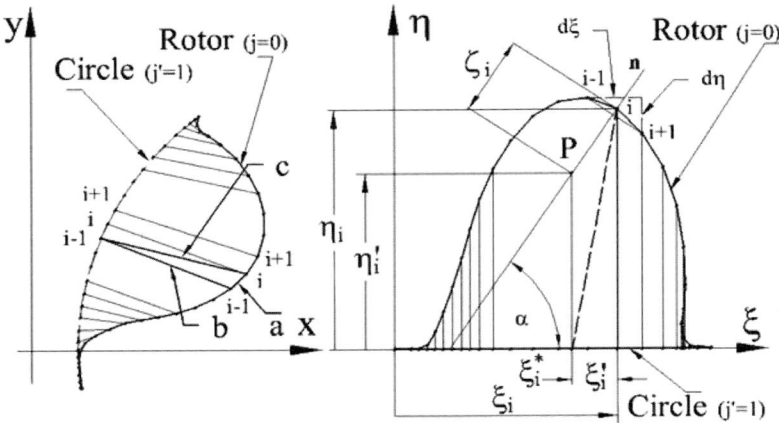

Figure 3-13 Transformation from physical (left) to computational domain (right)

The computational cells are formed between the points in the computational domain, as shown in Figure 3-13. As a consequence of the transformation, both the right and left cell boundaries are produced as vertical lines. Unfortunately, some of the cells generated by this procedure are either inverted or twisted. However, these can become regular if the points on the straight edge in the computational domain can be rearranged to form a monotonically increasing or decreasing sequence. In that case, the reverse transformation from the computational domain would also give a regular point distribution on boundaries in the physical domain.

Consider point i in the computational domain on the rotor boundary as shown in the right diagram of Figure 3-13. The line normal to the rotor profile at that point is defined by the angle between that line and the ξ axis as:

$$\tan \alpha = \frac{d\xi}{d\eta} \qquad (3.27)$$

Assume that point P lies on that normal line and its projection on the ξ axis gives the desired point distribution on the horizontal boundary, as shown in the diagram by the dashed line. The position of point P is defined by its vertical coordinate η'_i. The distance between points P and i is given by ζ_i, which is a function of the vertical coordinate of point P_i. In that case, the horizontal projection of the point P_i can be calculated as:

$$\xi^*_i = \xi_i - \xi'_i, \qquad (3.28)$$

The vertical projection is specified as:

$$\eta'_i = \eta_i - \zeta_i \cdot \sin \alpha. \qquad (3.29)$$

3.3 Generation and Adaptation of Domain Boundaries

The ratio between the distance ζ_i and the vertical coordinate of point P_i is given by:

$$k_i = \frac{\eta'_i}{\zeta_i}, \qquad (3.30)$$

where the coefficient k_i can be any number greater then or equal to zero. If k_i is equal to zero then the point P_i is positioned on the ξ axis. However, if $k \to \infty$ then ζ_i becomes zero and point P_i corresponds to point i on the rotor. In that case, the distribution of the points is unchanged. By inserting (3.30) into equations (3.28) and (3.29), the new point projection on the ξ axis becomes:

$$\xi_i^* = \xi_i - \eta_i \cdot \frac{\cos\alpha}{k_i + \sin\alpha}. \qquad (3.31)$$

If the coefficient k in equation (3.31) has a constant value $k=1$, the new point distribution in the computational domain is always regular, as shown in the left part of Figure 3-14.
It then remains to make the inverse transformation from the line in the computational domain to the circle in the physical domain with respect to the new arc-length:

$$\mathbf{r}^*_{i,j'=1} = \mathbf{r}^*_{i,j'=1}(x,y) = \mathbf{r}_{I,j'=1}\left(\frac{\xi_i^*}{\xi_I^*}\right). \qquad (3.32)$$

The result of that procedure is shown on the right diagram in Figure 3-14. Linear interpolation or extrapolation between corresponding points on the circle $j'=1$ and rotor $j=0$ gives the point on the rack, as shown in the right drawing of Figure 3-14.

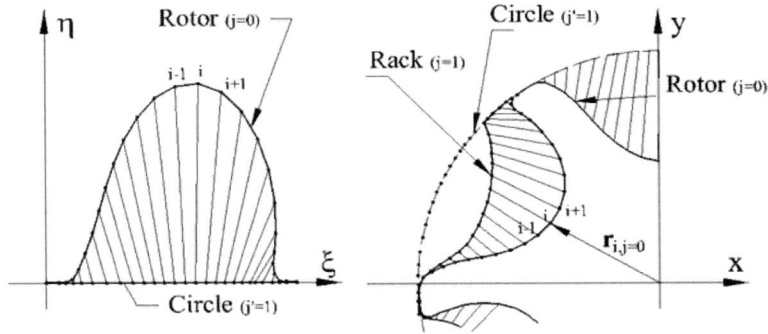

Figure 3-14 Final distribution of points in computational (left) and physical (right) domain

This method ensures the satisfactory distribution of boundary points in the 2D cross section of the screw machine rotor domains. The level of redistribution can be controlled by the factor k_i, which may be constant for all points through the domain or can be changed for each point by use of some characteristic parameter. For screw machine rotors, it appears that a constant value $k=1$ always gives a regular distribution of points on the boundaries. This is an essential prerequisite for successful generation of the internal points.

3.4 Algebraic Grid Generation for Complex Boundaries

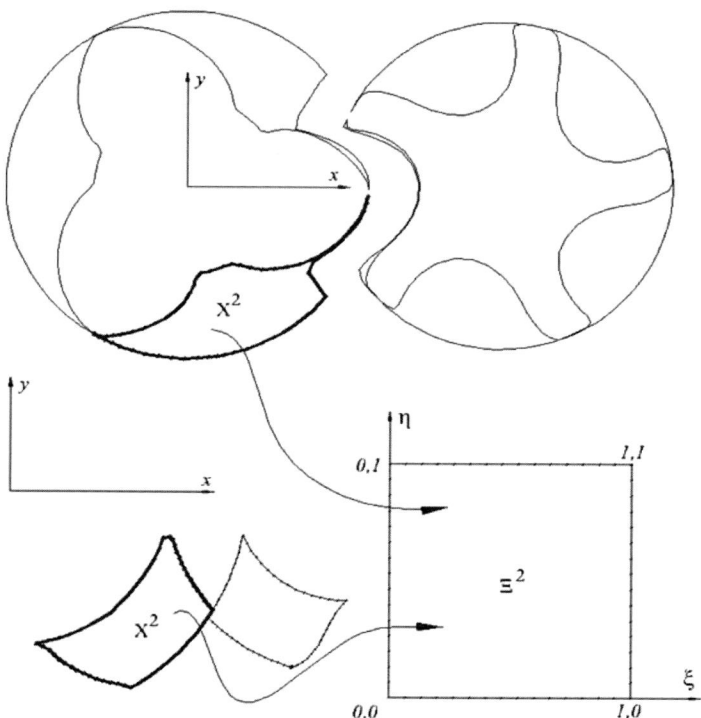

Figure 3-15 Conformal mapping of a physical domain X^2 to computational domain Ξ^2

Algebraic grid generation is often based on transfinite interpolation. This is defined as a multivariate interpolation procedure or a Boolean sum of univariate interpolations along each computational coordinate. The method is commonly used for grid generation in domains with smooth boundaries that are not highly deformed, or as an initial approximation for the iterative process of an elliptic grid solver. Transfinite interpolation, in conjunction with a boundary adaptation and orthogonalisation procedure, has been found to be a successful method for screw machine numerical grid generation and is therefore described here.

After the boundaries of a physical domain have been calculated and the boundary points adapted to the geometry conditions, they have to be mapped to a computational domain, in which the inner nodes of a computational mesh are estimated. The coordinates of the 2D physical domain are given in an x-y coordinate system while the computational coordinates are ξ-η as shown in Figure 3-15. The transfinite interpolation method is used for calculation of the inner point coordinates. Both, the block grid (left) and the "O" grid (right) are mapped on a similar computational grid. These domains are highlighted in the figure.

3.4.1 Standard Transfinite Interpolation

The coordinates of four boundary faces generated in two dimensions, with the help of adaptation methods, can be written in vector form as:

$$\mathbf{a}_l(\eta) = \mathbf{r}(\xi_l, \eta), \qquad l = 1,2 \qquad (3.33)$$
$$\mathbf{b}_l(\xi) = \mathbf{r}(\xi, \eta_l), \qquad l = 1,2 ,$$

where the coordinates of the transformed computational coordinate system, ξ and η, are:

$$\xi = \frac{(i-1)}{(I-1)} \text{ and } \eta = \frac{(j-1)}{(J-1)}.$$

i and j denote point numbers in physical coordinates while I and J are the overall number of points on these coordinates. As defined by the transfinite mapping method, the coordinates of the interior points are given as:

$$\mathbf{r}_1(\xi,\eta) = \sum_{l=1}^{2} \alpha_l(\xi)\mathbf{a}_l(\eta) \qquad (3.34)$$

$$\mathbf{r}(\xi,\eta) = \mathbf{r}_1(\xi,\eta) + \sum_{l=1}^{2} \beta_l(\eta)[\mathbf{b}_l(\xi) - \mathbf{r}_1(\xi,\eta_l)] ,$$

Blending functions $\alpha_1(\xi)$ and $\beta_1(\eta)$ are arbitrary functions of the computational coordinates that satisfy the cardinality conditions given by (3.35) to ensure that the edges of the domain are reproduced as a part of the solution.

$$\alpha_l(\xi_k) = \delta_{kl}, \qquad k=1,2 \quad l=1,2$$
$$\beta_l(\eta_k) = \delta_{kl}, \qquad k=1,2 \quad l=1,2 \qquad (3.35)$$

64   3 Grid generation of Screw Machine Geometry

where δ is the Kronecker delta.

By use of equation (3.33), equation (3.34) for a 2-D domain can be written in the following general form of the transfinite interpolation method. This connects the coordinates in the physical and numerical domains:

$$x(\xi,\eta) = X_1(\xi,\eta)\alpha_1(\xi) + X_2(\xi,\eta)\alpha_2(\xi)$$
$$y(\xi,\eta) = Y_1(\xi,\eta)\beta_1(\eta) + Y_2(\xi,\eta)\beta_2(\eta) \qquad (3.36)$$

The ability of the analytical transfinite interpolation method of (3.36) to produce a regular distribution of internal points is highly dependent upon the selection of the blending functions $\alpha_1(\xi)$ and $\beta_1(\eta)$. These functions define the curvature and orthogonality of the internal grid lines.

Lagrange Blending Functions

The simplest method of obtaining blending functions is by Lagrange interpolation:

$$\begin{array}{ll} \alpha_1(\xi) = 1-\xi & \alpha_2(\xi) = \xi \\ \beta_1(\eta) = 1-\eta & \beta_2(\eta) = \eta \end{array} \qquad (3.37)$$

Applying (3.37) to equation (3.36) gives the inner mesh points as:

$$x(\xi,\eta) = (1-\xi)X_1(\xi,\eta) + \xi X_2(\xi,\eta)$$
$$y(\xi,\eta) = (1-\eta)Y_1(\xi,\eta) + \eta Y_2(\xi,\eta) \qquad (3.38)$$

where X_1 and Y_1 are points on one boundary of a physical domain, while X_2 and Y_2 define the other boundary. The connections between the two opposite boundaries are produced here as straight lines, generally non-orthogonal to the boundaries. This method gives a satisfactory mesh only for simple geometries and is not usually applicable to the rotor domains of screw machines except as an initial grid for further orthogonalisation and smoothing. By this means, physical domains of a less complex shape can be successfully mapped, but orthogonality of the boundaries is not necessarily achieved. A numerical mesh generated by the transfinite interpolation method with Lagrange blending functions is presented in Figure 3-16. The discharge port of a compressor is mapped on the left of the figure, (a), while a working chamber with rotors of 3/5 configuration is presented on the right (b). The detail in the top left corner shows how the transition between the main domain and the clearances is mapped.

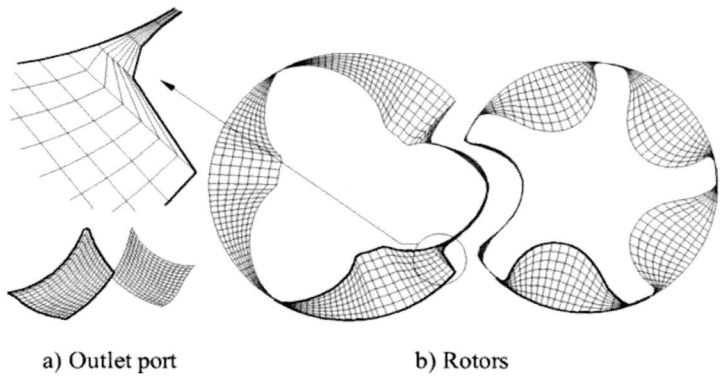

a) Outlet port   b) Rotors

Figure 3-16 Numerical mesh generated by TFI using Lagrange blending function

3.4.2 Ortho Transfinite Interpolation

A better and more accurate solution can be obtained by applying a modified transfinite interpolation formula with Hermite blending functions. Equation (3.34) can be written in a more general form as:

$$\mathbf{r}_1(\xi,\eta) = \sum_{l=1}^{2}\sum_{n=0}^{1} \alpha_l^n(\xi)\mathbf{a}_l^n(\eta) \tag{3.39}$$

$$\mathbf{r}(\xi,\eta) = \mathbf{r}_1(\xi,\eta) + \sum_{l=1}^{2}\sum_{n=0}^{1} \beta_l^n(\eta)[\mathbf{b}_l^n(\xi) - \frac{\partial^n}{\partial \eta^n}\mathbf{r}_1(\xi,\eta_l)],$$

and represents the Ortho transfinite interpolation formula. The representation of boundary points in this formula is similar to (3.33) but includes the derivatives on the boundaries:

$$\mathbf{a}_l^n(\eta) = \frac{\partial^n}{\partial \xi^n}\mathbf{r}(\xi_l,\eta), \quad\quad l=1,2 \quad n=0,1$$

$$\mathbf{b}_l^n(\xi) = \frac{\partial^n}{\partial \eta^n}\mathbf{r}(\xi,\eta_l), \quad\quad l=1,2 \quad n=0,1 \tag{3.40}$$

Equation (3.34) contains eight terms, four of which refer to the edges and four to the corners. On the other hand, equation (3.39) contains not only the terms for the coordinate data on four edges and four corners, but also their derivatives on four edges, two tangent vectors in each corner and a mixed derivative in each corner. This forms a combination of 24 terms.

Cardinality conditions are consequently more complex for ortho methods because the conditions are imposed on the slope of the blending function as:

66   3 Grid generation of Screw Machine Geometry

$$\frac{\partial^n}{\partial \xi^n} \alpha_l^n(\xi_k) = \delta_{kl} \delta_{nm}, \qquad k=1,2 \quad l=1,2 \quad n=0,1 \quad m=0,1$$

$$\frac{\partial^n}{\partial \eta^n} \beta_l^n(\eta_k) = \delta_{kl} \delta_{nm}, \qquad k=1,2 \quad l=1,2 \quad n=0,1 \quad m=0,1$$

(3.41)

Indices l and k in equations (3.39) to (3.41), correspond to the 2-D boundary faces, while the indices m and n indicate the corners.

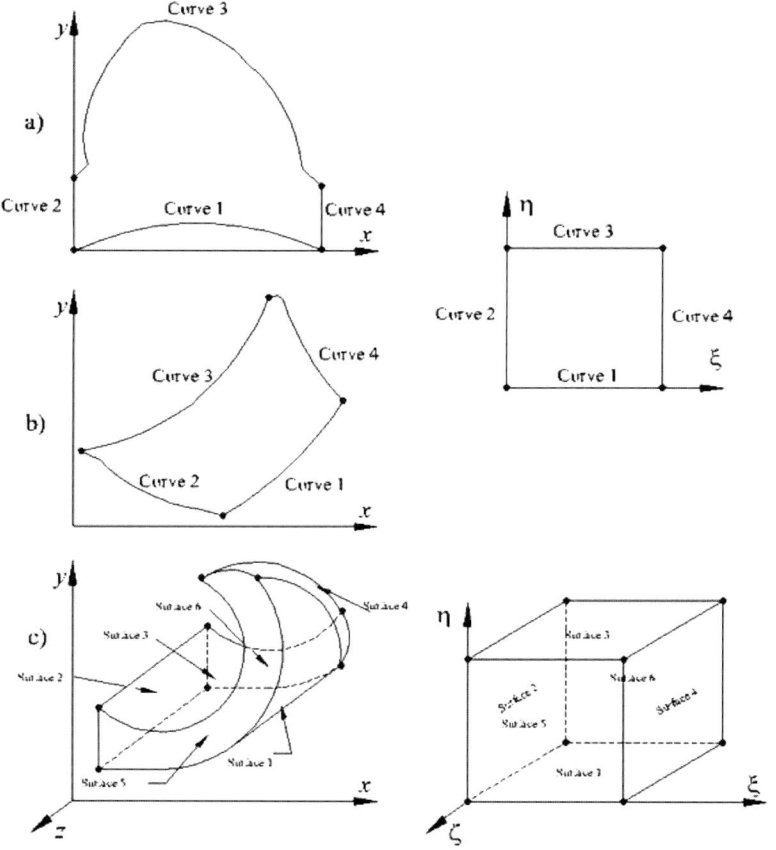

Figure 3-17 Classification of physical domains with respect to mapping requirements

Equation (3.39) applies to a two-dimensional domain however, it can easily be applied to three dimensions if the coordinate z is added perpendicularly to the x-y plane and accounted for in the equation. Physical domains, which have to be mapped, can be classified either as a two-, four- or six- boundary mapping prob-

3.4 Algebraic Grid Generation for Complex Boundaries

lem. In which group the problem is categorised depends not only on the dimensionality of the domain, but also on the characteristics of the boundaries. Examples are shown in Figure 3-17. Case a) is a two-boundary mapping problem where only two opposite boundaries in the y direction need to be mapped. In case b), all four boundaries must be mapped in order to preserve the regular boundary fitting. Case c) is a three-dimensional problem in which all six boundaries must be mapped. The geometry of the rotor domains in a screw machine can be regarded as 2½ dimensional. It is because the rotors of a screw machine are generated from a profile which is revolved around its axis and moved along the same axis. In that case, the four boundary mapping method has to be applied at each of the 2-D cross sections, which are later connected to obtain a three dimensional mesh.

Hermite Blending Functions

Contrary to the Lagrange blending functions used in the previous section, Hermite blending functions involve derivatives at the end curve points to enforce orthogonality at the boundaries. The blending functions are applied in a cubic form as:

$$\alpha_1^0 = h_1(\eta) = 2\eta^3 - 3\eta^2 + 1, \quad \alpha_1^1 = h_2(\eta) = -2\eta^3 + 3\eta^2, \quad (3.42)$$
$$\alpha_2^0 = h_3(\eta) = \eta^3 - 2\eta^2 + \eta, \quad \alpha_2^1 = h_4(\eta) = \eta^3 - \eta^2$$

which satisfy cardinality conditions (3.41). The Hermite blending functions h_1 to h_4, defined by the previous equation, are now functions only of the computational coordinate η. Applying these functions to equation (3.39) for a two dimensional problem, one can easily get:

$$x'(\xi,\eta) = X_1(\xi)h_1(\eta) + X_2(\xi)h_2(\eta) + \frac{\partial x(\xi,\eta_0)}{\partial \eta}h_3(\eta) + \frac{\partial x(\xi,\eta_1)}{\partial \eta}h_4(\eta)$$
$$y'(\xi,\eta) = Y_1(\xi)h_1(\eta) + Y_2(\xi)h_2(\eta) + \frac{\partial y(\xi,\eta_0)}{\partial \eta}h_3(\eta) + \frac{\partial y(\xi,\eta_1)}{\partial \eta}h_4(\eta) \quad (3.43)$$

The values for computational coordinates at the start and end points in the previous equation are $\eta_0 = 0$ and $\eta_1 = 1$. The boundary points in the same equation are

$$X_1 = x(\xi,\eta_0) = X_1(\xi) \qquad Y_1 = y(\xi,\eta_0) = Y_1(\xi)$$
$$X_2 = x(\xi,\eta_1) = X_2(\xi) \qquad Y_2 = y(\xi,\eta_1) = Y_2(\xi) \quad (3.44)$$

Equations (3.43) represent a two-boundary method of transfinite interpolation procedure with Hermite blending functions for two-dimensional domains. The partial derivatives at the boundaries ensure orthogonality. These are:

$$\frac{\partial x(\xi,\eta_0)}{\partial \eta} = K_1(\xi)(-\frac{\partial Y_1}{\partial \xi}) \qquad \frac{\partial x(\xi,\eta_1)}{\partial \eta} = K_2(\xi)(-\frac{\partial Y_2}{\partial \xi})$$
$$\frac{\partial y(\xi,\eta_0)}{\partial \eta} = -K_1(\xi)(-\frac{\partial X_1}{\partial \xi}) \qquad \frac{\partial y(\xi,\eta_1)}{\partial \eta} = -K_2(\xi)(-\frac{\partial X_2}{\partial \xi}) \qquad (3.45)$$

The coefficients K_1 and K_2 are positive numbers smaller than 1. They are usually chosen by trial and error to avoid overlapping of connecting curves inside the domain.

The four-boundary method, which assumes interpolation between all four boundaries in the 2-D domain can be written in the following form:

$$x(\xi,\eta) = x'(\xi,\eta) + \Delta x(\xi,\eta)$$
$$y(\xi,\eta) = y'(\xi,\eta) + \Delta y(\xi,\eta) \qquad (3.46)$$

For this boundary method, two additional opposite boundaries must be mapped. These are:

$$X_3 = x(\xi_0,\eta) = X_3(\eta), \; Y_3 = y(\xi_0,\eta) = Y_3(\eta)$$
$$X_4 = x(\xi_1,\eta) = X_4(\eta), \; Y_4 = y(\xi_1,\eta) = Y_4(\eta) \qquad (3.47)$$

The first term in (3.46) is calculated in equation (3.43) while the second term defines a mapping between the other two boundaries as:

$$\Delta x(\xi,\eta) = (X_3 - X'_3)h_5(\xi) + (X_4 - X'_4)h_6(\xi) + \left(\frac{\partial x(\xi_0,\eta)}{\partial \xi} - \frac{\partial x'(\xi_0,\eta)}{\partial \xi}\right)h_7(\xi)$$
$$+ \left(\frac{\partial x(\xi_1,\eta)}{\partial \xi} - \frac{\partial x'(\xi_1,\eta)}{\partial \xi}\right)h_8(\xi) \qquad (3.48)$$
$$\Delta y(\xi,\eta) = (Y_3 - Y'_3)h_5(\xi) + (Y_4 - Y'_4)h_6(\xi) + \left(\frac{\partial y(\xi_0,\eta)}{\partial \xi} - \frac{\partial y'(\xi_0,\eta)}{\partial \xi}\right)h_7(\xi)$$
$$+ \left(\frac{\partial y(\xi_1,\eta)}{\partial \xi} - \frac{\partial y'(\xi_1,\eta)}{\partial \xi}\right)h_8(\xi)$$

The partial derivatives at the boundary points in equation (3.48) ensure that appropriate derivatives in the corners of the domain are accounted as part of the solution:

3.4 Algebraic Grid Generation for Complex Boundaries

$$\frac{\partial x'(\xi_0,\eta)}{\partial \xi} = h_1(\eta)\frac{\partial x(\xi_0,\eta_0)}{\partial \xi} + h_2(\eta)\frac{\partial x(\xi_0,\eta_1)}{\partial \xi}$$
$$\frac{\partial x'(\xi_1,\eta)}{\partial \xi} = h_1(\eta)\frac{\partial x(\xi_1,\eta_0)}{\partial \xi} + h_2(\eta)\frac{\partial x(\xi_1,\eta_1)}{\partial \xi} \quad , \tag{3.49}$$
$$\frac{\partial y'(\xi_0,\eta)}{\partial \xi} = h_1(\eta)\frac{\partial y(\xi_0,\eta_0)}{\partial \xi} + h_2(\eta)\frac{\partial y(\xi_0,\eta_1)}{\partial \xi}$$
$$\frac{\partial y'(\xi_1,\eta)}{\partial \xi} = h_1(\eta)\frac{\partial x(\xi_1,\eta_0)}{\partial \xi} + h_2(\eta)\frac{\partial x(\xi_1,\eta_1)}{\partial \xi}$$

while the remaining partial derivatives in the same equation account for derivatives on two additional edges as:

$$\frac{\partial x(\xi_0,\eta)}{\partial \xi} = -K_3(\eta)(-\frac{\partial Y_3}{\partial \eta}) \qquad \frac{\partial x(\xi_1,\eta)}{\partial \xi} = -K_4(\eta)(-\frac{\partial Y_4}{\partial \eta})$$
$$\frac{\partial y(\xi_0,\eta)}{\partial \xi} = K_3(\eta)(-\frac{\partial X_3}{\partial \eta}) \qquad \frac{\partial y(\xi_1,\eta)}{\partial \xi} = K_4(\eta)(-\frac{\partial X_4}{\partial \eta}) \tag{3.50}$$

The remaining Hermite factors are:

$$h_5 = 2\xi^3 - 3\xi^2 + 1, \qquad h_6 = -2\xi^3 + 3\xi^2$$
$$h_7 = \xi^3 - 2\xi^2 + \xi, \qquad h_8 = \xi^3 - \xi^2 \tag{3.51}$$

The four-boundary Hermite interpolation method gives reasonably good distribution of internal points with freedom to maintain the orthogonality and curvature on and near the boundaries. This method usually gives a sufficiently good result for the domains of the screw machine inlet and outlet ports. However, for the rotor domains of a screw machine, where the geometry changes rapidly, it sometimes causes the internal domain lines to overlap or even to exceed the boundaries. The intensity of that depends on the values selected for the coefficients K_1 to K_4. Figure 3-18 shows some domains, as in Figure 3-16, but this time the four-boundary method for Ortho transfinite interpolation is used in combination with Hermite blending functions to generate the grid. On rotors, in the part where the radial distance between the opposite sides changes rapidly, this method results in overlap and even in grid lines exceeding the boundary lines. In other parts of the mesh, the result is sufficiently good and the internal lines are orthogonal to the boundaries. At the outlet port, which is significantly easier to map, the resulting numerical mesh is regular, boundary fitted and boundary orthogonal. The numerical mesh for the rotor domains presented in this figure is generated with very low values of coefficients K_1 to K_4.

70   3 Grid generation of Screw Machine Geometry

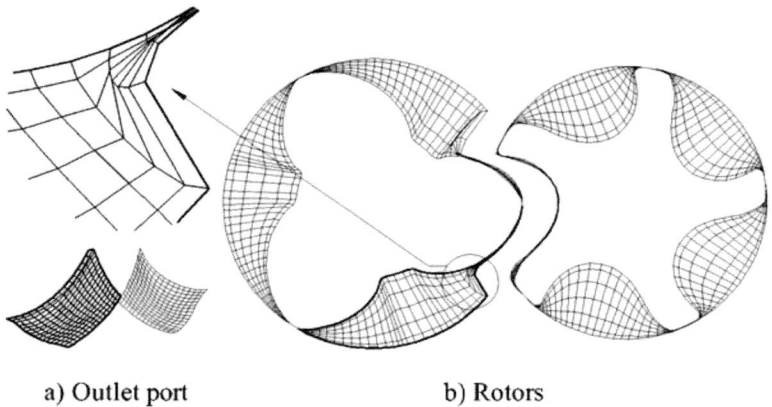

a) Outlet port   b) Rotors

Figure 3-18 Numerical mesh generated by TFI using Hermite blending function

Multidimensional Stretching Functions

The algebraic methods presented in previous sections employ either Lagrange or Hermite blending functions to control the distribution of grid points in the physical domain. These functions are generated from the one-dimensional stretching functions ξ and η with equal values for all grid lines throughout the domain. Since this is a limitation for complex geometries, another method is introduced based on the construction of multidimensional stretching functions as proposed by *Steinthorrsson et al (1992)*. A two-dimensional grid system with x and y coordinates in the physical domain and with coordinates ξ and η in the computational domain is analysed. In the transformed computational domain, coordinates ξ and η vary between zero and one as shown in Figure 3-15. Considering the coordinate ξ it is obvious that all points on boundaries ξ_o and ξ_l affect the distribution of the inner points in that direction. The same situation appears for the η coordinates. Therefore, two one-dimensional stretching functions in ξ direction can be identified for two opposite boundaries: $\hat{\xi}_0(\xi), \hat{\xi}_1(\xi)$ for the points on $\eta=0$ and $\eta=1$ respectively. The edges of the computational domain are one-dimensional lines on which one-dimensional stretching functions can be easily obtained either by Lagrange (3.37) or by Hermite interpolation (3.51). In this study, the point distribution on the edges is known and therefore the stretching functions at the boundaries are known in advance.

Multidimensional stretching functions can be determined once the one-dimensional functions are selected. By application of linear Lagrange interpolation, the following multidimensional function can be constructed:

$$\hat{\xi}(\xi,\eta) = \hat{\xi}_0(\xi) \cdot (1-\eta) + \hat{\xi}_1(\xi) \cdot \eta . \qquad (3.52)$$

3.4 Algebraic Grid Generation for Complex Boundaries

To improve the situation, cubic polynomials similar to Hermite interpolation factors can be used in order to produce multidimensional stretching functions of the following form:

$$\widehat{\xi}(\xi,\eta) = \widehat{\xi}_0(\xi) \cdot h_1(\eta) + \widehat{\xi}_1(\xi) \cdot h_2(\eta), \tag{3.53}$$

where coefficients h_1 and h_2 are obtained from equation (3.42). These coefficients can be functions not only of the computational coordinate η, but also they can be specified as a function of the arc-length of the newly calculated inner grid line, i.e. its natural coordinate, as:

$$h_1(s) = 2s^3 - 3s^2 + 1, \quad h_2(s) = -2s^3 + 3s^2, \tag{3.54}$$

If the boundaries of the physical domain have to be mapped together with the inner points, then analytical stretching functions have to be calculated each time a boundary is mapped and then updated throughout the entire domain. However, if the boundary points are distributed on the edges in advance, the procedure looks significantly simpler because the natural coordinate s defines a stretching function directly through its arc-length. Based on that, the equivalent stretching functions can be defined as:

$$\begin{aligned} &\widehat{\xi}_0(\xi_i) = 0 &&\text{for} \quad i = 1 \\ &\widehat{\xi}_0(\xi_i) = \frac{d_i}{d_L} &&\text{for} \quad i = 2, 3, \ldots L \end{aligned} \tag{3.55}$$

where

$$d_i = \sum_{n=2}^{i} \sqrt{(x_n - x_{n-1})^2 + (y_n - y_{n-1})^2} \qquad d_L = \sum_{i=1}^{L} d_i, \tag{3.56}$$

They are used to construct multidimensional stretching functions in (3.52) or (3.53).

The method described above can be applied on all, one-, two- or three-boundary procedures simply by replacing the one-dimensional stretching functions with the multidimensional stretching functions (3.53). This method often gives a satisfactory grid quality for complex systems and it is not too difficult to apply. However, in certain points of the mesh, where gradients of the cell size are extremely high, this method can produce instabilities and severe irregularities of the mesh, as a consequence of attempts to keep the mesh orthogonal. The numerical mesh generated by this method is shown in Figure 3-19. The error is magnified 100 times to become visible. The other parts of the mesh are mapped regularly with satisfactory control on orthogonality and smoothness. To use this method further, one must try to prevent unwanted oscillations.

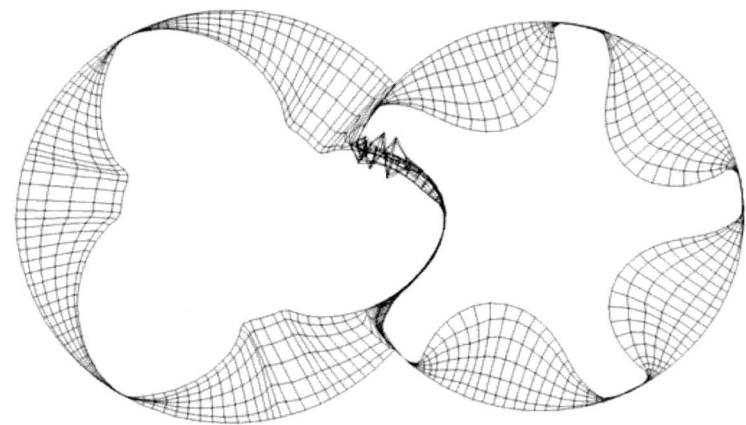

Figure 3-19 Numerical mesh generated with multidimensional stretching functions, where errors are magnified 100 times

Blending Functions Based on a Tension Spline Interpolation

Hermite transfinite interpolation based on cubic polynomials makes possible extensive skewness or overlap of the grid lines. This is the result of the use of K-factors in the equations. The reason is that K factors control the magnitude of the first-derivatives, which in turn control the curvature of the lines. The curvature of these lines is also affected by normal vectors on the edge. It is possible to correct these factors by implementing normalised boundary vectors. Although this approach is simple for a two-boundary method, it requires special attention in the case of a four-boundary method, because of the constraint on orthogonality in the interior and boundary points.

Similar control of the curvature of lines can be achieved if blending functions based on the tension spline interpolation are used instead of Hermite blending functions. The tension spline interpolation changes its nature with change in the tension parameter σ, which was first introduced by *Thompson et al* (1985).

To use blending functions based on tension spline interpolation, the Hermite blending factors in equation (3.46) should be replaced by:

$$h_1(s) = c_1(1-s) + c_2 s + c_2 \left[\frac{\sinh[(1-s)\sigma] - \sinh(s\sigma)}{\sinh(\sigma)} \right]$$

$$h_2(s) = c_1 s + c_2(1-s) - c_2 \left[\frac{\sinh[(1-s)\sigma] - \sinh(s\sigma)}{\sinh(\sigma)} \right] , \qquad (3.57)$$

$$h_3(s) = c_3 \left[(1-s) - \frac{\sinh[(1-s)\sigma]}{\sinh(\sigma)} \right] + c_4 \left[s - \frac{\sinh(s\sigma)}{\sinh(\sigma)} \right]$$

$$h_4(s) = -c_3 \left[s - \frac{\sinh(s\sigma)}{\sinh(\sigma)} \right] - c_4 \left[(1-s) - \frac{\sinh[(1-s)\sigma]}{\sinh(\sigma)} \right]$$

where the coefficients are defined as:

$$c_1 = 1 - c_2, \qquad c_2 = \frac{\sinh(\sigma)}{2\sinh(\sigma) - \sigma \cosh(\sigma) - \sigma}$$

$$c_3 = \frac{-\alpha}{(\beta^2 - \alpha^2)} \sinh(\sigma), \qquad c_4 = \frac{\beta}{(\beta^2 - \alpha^2)} \sinh(\sigma) \qquad (3.58)$$

$$\alpha = \sigma \cosh(\sigma) - \sinh(\sigma), \qquad \beta = \sinh(\sigma) - \sigma$$

The curvature of grid lines is now controlled by the tension parameter σ. If the tension parameter increases to infinity ($\sigma \to \infty$), blending functions $h_1(s) \to (1-s)$, $h_2(s) \to s$, $h_3(s) \to 0$ and $h_4(s) \to 0$, based on the tension spline interpolation, become linear. However, decrease in this factor to zero ($\sigma \to 0$) causes the blending functions to become cubic polynomials. This feature gives significant flexibility to the method. Therefore, equation (3.57) can be used to interpolate any function in the interval (0,1) of the computational coordinate if the function values and first derivatives are known at the end points of the interval. The flexibility of this method ensured its significant role in the grid generation of screw machines.

3.4.3 Simple Unidirectional Interpolation

There are cases where the complex, time and space consuming approach presented in the previous section is not essential for achieving satisfactory meshes. This is especially true for regions of less-complex boundary shapes, when boundary orthogonalisation is required, or for initial meshes for more complex domains. This is the case of the outlet port or oil injection port, for example. The idea is to use a simple unidirectional interpolation instead of a transfinite interpolation, as proposed by *Zhou* (1998). Interpolation is applied between two opposite, non-contacting boundaries of a two-dimensional domain. The method, although simple, can be easily compared with the two-boundary transfinite interpolation method, because its nature is originally two-dimensional. However, the same method can be applied to all four boundaries of the domain if unidirectional interpolation is applied in turn to both pairs of boundaries.

Consider the case in Figure 3-15 in which a physical domain in the *x-y* coordinate system has to be mapped by the use of a computational domain in the ξ-η coordinate system. Both the ξ and η coordinates of the computational domain vary between 0 and 1. Uni-directional interpolation in the following mathematical expression is applied to obtain the grid points of the physical domain:

$$x(\xi,\eta) = x(\xi_0,\eta) + \beta_\xi \left[x(\xi_1,\eta) - x(\xi_0,\eta) \right]$$
$$y(\xi,\eta) = y(\xi,\eta_0) + \beta_\eta \left[y(\xi,\eta_1) - y(\xi,\eta_0) \right] \tag{3.59}$$

The stretching functions β_ξ and β_η are defined by Lagrange interpolation as:

$$\beta_\xi = (1-w_\eta)\alpha_1(\xi) + w_\eta \alpha_0(\xi)$$
$$\beta_\eta = (1-w_\xi)\alpha_1(\eta) + w_\xi \alpha_0(\eta) \tag{3.60}$$

where w_ξ and w_η are weight functions of the appropriate coordinates and of the constant *m*. The aim of this constant is to control the effect of boundary stretching functions α_1 and α_0. The weight functions are:

$$w_\xi = \frac{(1-\xi)^m}{\xi^m + (1-\xi)^m}; \qquad w_\eta = \frac{(1-\eta)^m}{\eta^m + (1-\eta)^m}. \tag{3.61}$$

The exponent *m* in the above equations takes the values of 1, 2 or 3 which correspond to linear, quadratic and cubic transformation between the opposite boundaries respectively. Stretching functions on the boundaries, which are implemented in equation (3.60), are calculated as a projection of the appropriate computational boundary on the *x* or *y* axis. Which axis is used for the projection depends on the difference obtained between the two end points on that boundary. If the difference between the end points on the original physical coordinate differs from zero, then that coordinate is used. The specification is given with:

$$\alpha_K(\xi) = \begin{cases} \dfrac{x(\xi,K) - x(0,K)}{x(1,K) - x(0,K)} & \text{if } \left[x(1,K) - x(0,K) \right] \neq 0 \\ \dfrac{y(\xi,K) - y(0,K)}{y(1,K) - y(0,K)} & \text{if } \left[x(1,K) - x(0,K) \right] = 0 \end{cases} \quad K=0,1 \tag{3.62}$$

and

3.4 Algebraic Grid Generation for Complex Boundaries   75

$$\alpha_K(\eta) = \begin{cases} \dfrac{y(K,\eta)-y(K,0)}{y(K,1)-y(K,0)} & \text{if } \left[y(K,1)-y(K,0)\right] \neq 0 \\ \dfrac{x(K,\eta)-x(K,0)}{x(K,1)-x(K,0)} & \text{if } \left[y(K,1)-y(K,0)\right] = 0 \end{cases} \quad K=0,1 \quad (3.63)$$

Examples of use of a simple unidirectional interpolation are given in Figure 3-20.

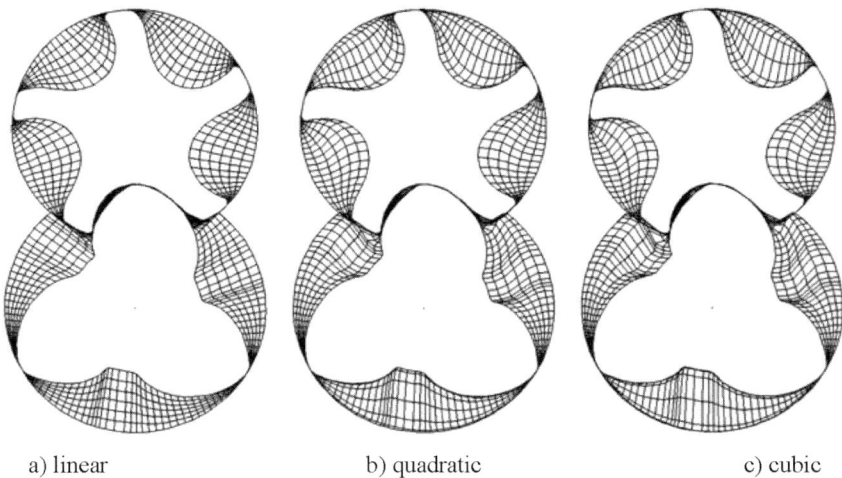

a) linear　　　　　　　　　b) quadratic　　　　　　　　　c) cubic

Figure 3-20 Numerical meshes generated by simple unidirectional interpolation

The use of a linear weight function is presented in Figure 3-20a. The final mesh is very similar to the one generated by a Lagrange TFI and even the effort required to obtain the result is almost the same. A quadratic weight function Figure 3-20b gives a better result for simple numerical meshes. However, for complex meshes it usually gives overlapping grid lines in regions where the mesh changes rapidly. More importantly, the generation of derivatives on boundaries depends on the absolute position of the boundary in the physical domain. This causes unwanted and unphysical curvature of the inner grid lines. Therefore, it appears to be almost unacceptable for the generation of complicated screw compressor grids. A similar situation occurs with a cubic weight function Figure 3-20c, which produces even higher uncertainties about the mesh.

3.4.4 Grid Orthogonalisation

By combining the analytical grid generation methods mentioned above, it should generally be possible to produce a satisfactory grid for the CFD analysis of screw machine flows. If this is not the case, as for rotors with a very small radius on the lobe tips, orthogonalisation and smoothing, together with boundary adaptation must be applied. The approach to orthogonalisation of a screw compressor grid is similar to one suggested by *Lehtimaki* (2000).

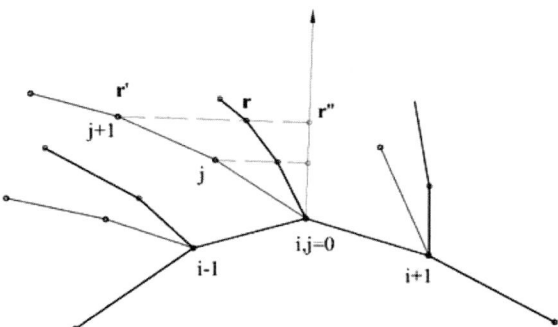

Figure 3-21 Orthogonalisation to the boundary line j=0

Ortho transfinite interpolation with Hermite blending factors provides inherent orthogonality properties. However, this is controlled by *K*-factors on boundaries which in certain cases lead to overlapping of the grid lines. The orthogonalisation proposed here is, however, independent of the grid generation process and, therefore, can be performed on any grid generated in advance. This can lead to significant savings in the computational effort otherwise wasted in generation of an orthogonal grid. The method is based on the moving of the computational point towards its orthogonal projection normal to the boundary, as presented in Figure 3-21.

The position of a numerical point in the physical domain, calculated by some algebraic grid generating method, which has to be orthogonalised, is given by vector $\mathbf{r'}_{i,j}$ while the perpendicular projection of that point to the normal of the boundary is $\mathbf{r''}_{i,j}$. A weighting factor between the original point $\mathbf{r'}_{i,j}$ and the projected point is applied to avoid over specification caused by any discrepancy between the two boundaries. The new point is therefore defined by its radius vector:

$$\mathbf{r}_{i,j} = (1 - w_{i,j})\mathbf{r'}_{i,j} + w_{i,j}\mathbf{r''}_{i,j} \qquad (3.64)$$

The weighting factor has an exponential form:

$$w_{i,j} = \exp\left\{-C_1\left[\left(1 - \frac{\tilde{\eta}_{i,j}}{\tilde{\eta}_{i,j_{max}}}\right) - 1\right]\right\} \cdot \left[4\xi_{i,0}(1-\xi_{i,0})\right]^{C_2} \qquad (3.65)$$

where $\tilde{\eta}_{i,j} = \sqrt{(\xi_{i,j} - \xi_{i,0})^2 + (\eta_{i,j} - \eta_{i,0})^2}$ are the arc-length values. The first term in the equation is used for damping of the interior of the grid while the second affects damping at the boundary line. Coefficients are provided to control the amount of damping. Both coefficients in equation (3.65) should have positive values. Higher values of C_1 make damping of the orthogonalisation to the interior of the domain higher, while a higher C_2 reduces the region which is orthogonalised to the central part of the boundary. The intensity of orthogonalisation is always damped to a certain level defined by the coefficients C_1 and C_2. It implies that the orthogonalisation procedure can be applied many times in succession. Each repetition leads to further orthogonalisation of the mesh.

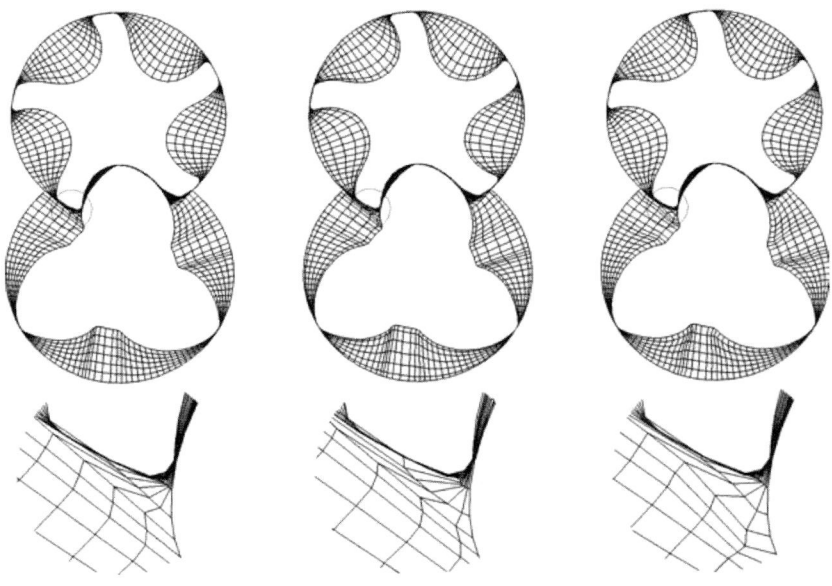

a) Lagrange Standard TFI   b) Hermite Ortho TFI   c) Orthogonalisation
Figure 3-22 Comparison of different grid generation methods

Three numerical meshes are compared in Figure 3-24. Case a) is a standard TFI with Lagrange blending functions which gives a regular but non-orthogonal mesh. This mesh is later used as the basis for the orthogonalisation process. The middle case b) is Ortho TFI with Hermite blending functions. The K factors are set to 0.2 to avoid overlapping of grid lines. Therefore, the lines in the main domain tend to be linear, while in the clearance gaps these become more curved but, often, not properly controlled. In the third case c), the mesh generated by standard TFI (case a) is additionally orthogonalised. The result achieved by orthogonalisation in c) is better than both the previous cases. Although the damping of orthogonalisation of grid lines is set to be high, the mesh is orthogonal in both the main domains and the gaps. The mesh is more consistent, and regular with lower values of aspect and expansion ratios than in both the previous cases.

Because of its obvious advantages, this method is always applied for the generation of the final mesh used for finite volume calculation of fluid flow in screw machinery. It can be even further improved by smoothing the numerical mesh, as is described in the next section.

3.4.5 Grid Smoothing

Discontinuities which appear at the mesh boundary, despite the mesh being orthogonalised, propagate in the interior of the domain, causing cells along these lines to remain non orthogonal. The problem can be conveniently solved by introducing a smoothing procedure. The following formula is among the easiest to apply:

$$x_{i,j}^{n+1} = x_{i,j}^n + C\left(x_{i+1,j}^n - 2x_{i,j}^n + x_{i-1,j}^n\right)$$
$$y_{i,j}^{n+1} = y_{i,j}^n + C\left(y_{i+1,j}^n - 2y_{i,j}^n + y_{i-1,j}^n\right)$$
(3.66)

where C is constant and $n=0,1,2,\ldots,N$ is the number of repetitions. This constant must satisfy condition $C \leq 0.5$ to preserve the stability of the method.

Equation (3.66) can be repeatedly applied as many times as required to obtain a smooth grid. However, its use can affect orthogonality on the boundaries and, therefore, it must be used with care. The difference between a mesh to which smoothing is not applied (left) and a smooth grid is shown in Figure 3-23.

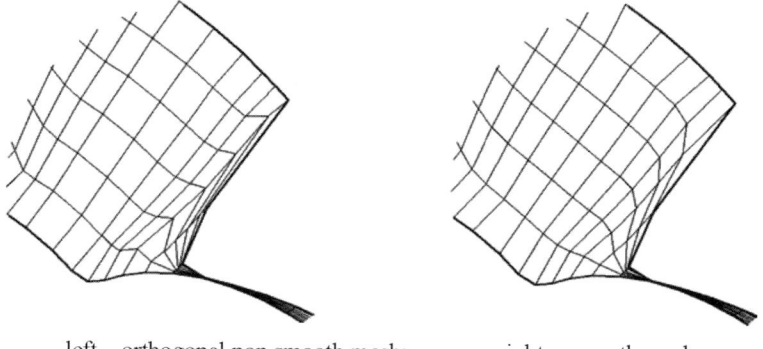

left – orthogonal non smooth mesh;   right – smooth mesh

Figure 3-23 Application of smoothing function

3.4.6 Moving Grid

A numerical mesh generated by methods presented previously consists of vertex, cell and region definitions. Each vertex is uniquely defined by its coordinate in an absolute coordinate system:

$$\mathbf{r}_i = \mathbf{r}_i(x, y, z, \tau), \qquad (3.67)$$

and the vertex number V_i. Control volumes, i.e. computational cells are generally hexahedral, defined by eight vertices. These can also be deformed by merging vertices, Figure 3-9.

Rotor movement in a screw machine is rotation around the rotor axis. The male and female rotors rotate in opposite directions to each other at angular speeds proportional to the number of lobes z_1/z_2.

To simulate rotation of the rotors by movement of the numerical mesh, one has only to define a new vertex position before starting the calculation for each time step. By this means, the cells are moved and deformed without any redefinition other than specifying new positions of the vertices. Although, the inner boundary of the mesh, which belongs to a rotor, rotates around the axis, the other one, represented by the rack and housing does not rotate. The imaginary rack plane used to divide the working chamber of the screw machine rotors in two parts is visible on the main rotor domain in Figure 3-24. It is connected to the housing bore of the machine along a straight line where the bores of the male and female rotors connect. The rack plane moves in translation with the speed of the male rotor pitch circle. However, the other part of the same boundary, which represents the machine housing, rotates at the same angular speed as the rotor. Therefore, a complex combination of rotation and translation of the outer boundary has to be performed. It seems almost impossible to do that without disturbing the consistency of the mesh. Therefore, another approach is used.

The rotors of a screw machine are specially profiled helical gears. These are completely specified by the profile coordinates, rotor length and helix angle. The coordinates of the profile explicitly define the rotor diameter. One can imagine that the movement of the rotor domain can be simulated by removing the last cross section segment on one side of the rotor, when all the remaining vertices are moved axially along the axis such that only their z coordinate is changed and finally an appropriate segment is added to the other side of the rotor. By this means, the rotor is mapped in the following time step with the same number of vertices whose positions are only redefined, while the cell definitions remain unchanged. This procedure is simple and reliable but an additional file, containing vertices with an extended number of cross sections, has to be generated.

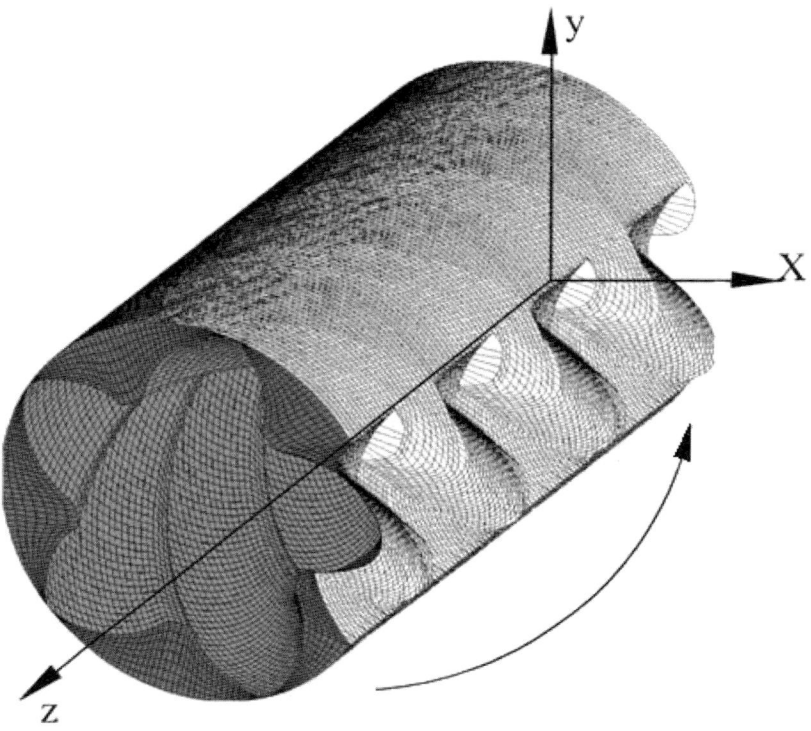

Figure 3-24 Moving strategy for screw machine numerical mesh

In Figure 3-24, a mesh which consists of a number of cross sections along the z axis is shown. Cross sections are defined at equal distances from each other

$$\Delta z = L / {z_1 \, n_{lobe}},$$

where L is the rotor length, z_1 is the number of teeth on the male rotor and n_{lobe} is the number of cross sections for one interlobe distance on the z axis. By this means, the cell length in the axial direction is specified as constant for all the cells in the domain. This has great advantages for the grid movement, but its weak point is that the time step and machine rotational speed are directly coupled. To show that, a unit angle is defined as

$$\delta\alpha = \frac{\phi_h}{z_1 \, n_{lobe}}, \tag{3.68}$$

which defines rotation between the two time steps, while the time step is

$$\delta t = \frac{60 \, \delta \alpha}{2 \pi n}, \tag{3.69}$$

where n is the speed of rotation, (rpm). Therefore, the time step is inversely proportional to the speed of rotation. That, in turn, means that the numerical mesh has to be changed for different rotation speeds in order to keep the ratio of time and spatial step constant. It's not always essential to keep the ratio constant but it is certainly limited by Courant stability conditions which give a maximum allowed ratio of time and spatial discretisation as:

$$\frac{|u \pm c| \Delta t}{\Delta x} < \alpha_c \tag{3.70}$$

where c is the speed of sound and α_c is a parameter which according to *Peric* (1990) depends on the particular time advance method used.

3.5 Computer Program

The procedures described in this chapter have been employed to form a stand alone CAD-CFD interface to generate a 3-D mesh of a screw machine working domain. The interface program is written in Fortran and is named SCORG, which stands for **S**crew **CO**mpressor **R**otor grid **G**enerator. The program calculates a numerical mesh for a screw machine based on given rack or rotor curves, by means of boundary adaptation and transfinite interpolation for all domains within the screw machine. These are: the working chamber, which surrounds the rotors and is divided into two parts, of which one belongs to each male and female rotor, the inlet and outlet ports and other openings, which may be pre-specified, like an oil injection port, or specified by a user program.

Transfer files are produced and imported into a numerical solver. Separate files are produced for the node, cell and region definitions for each domain of screw machine geometry. The transfer files also contain user subroutines for the numerical solver. These specify the initial and boundary conditions, the grid movement, the control parameters and the post-processing functions. The interface files are imported into a commercial CFD package through its pre-processor.

4

Applications

4.1 Introduction

This chapter demonstrates the scope of the method developed for the three-dimensional analysis of a screw compressor. The CFD package used in this case was COMET developed by ICCM GmbH Hamburg, today a part of CD-Adapco. The analysis of the flow and performance characteristics of a number of types of screw machines is performed to demonstrate a variety of parameters used for grid generation and calculation.

The first example is concerned with a dry air screw compressor. A common compressor casing is used with two alternative pairs of rotors. The rotors have identical overall geometric properties but different lobe profiles. The application of the adaptation technique enables convenient grid generation for geometrically different rotors. The results obtained by three dimensional modelling are compared with those derived from a one-dimensional model, previously verified by comparison with experimental data. The relative advantages of each rotor profile are demonstrated.

The second example shows the application of three dimensional flow analysis to the simulation of an oil injected air compressor. The results, thus obtained, are compared with test results obtained by the authors from a compressor and test rig, designed and built at City University. They are presented in the form of both integral parameters and a p-α indicator diagram. Calculations based on the assumptions of the laminar flow are compared to those of turbulent flow. The effect of grid size on the results is also considered and shown here.

The third example gives the analysis of an oil injected compressor in an ammonia refrigeration plant. This utilises the real fluid property subroutines in the process calculations and demonstrates the blow hole area and the leakage flow through the compressor clearances.

The fourth example presents two cases, one of a dry screw compressor to show the influence of thermal expansion of the rotor on screw compressor performance and one of a high pressure oil-flooded screw compressor to show the influence of high pressure loads upon the compressor performance.

4.2 Flow in a Dry Screw Compressor

Dry screw compressors are commonly used to produce pressurised air, free of any oil. A typical example of such a machine, similar in configuration to the compressor modelled, is shown in Figure 4-1. This is a single stage machine with 4 male and 6 female rotor lobes. The male and female rotor outer diameters are 142.380 mm and 135.820 mm respectively, while their centre lines are 108.4 mm apart. The rotor length to main diameter ratio l/d=1.77. Thus, the rotor length is 252.0 mm. The male rotor with wrap angle α_w=248.4^0 is driven at a speed of 6000 rpm by an electric motor through a gearbox. The male and female rotors are synchronised through timing gears with the same ratio as that of the compressor rotor lobes i.e. 1.5. The female rotor speed is therefore 4000 rpm. The male rotor tip speed is then 44.7m/s, which is a relatively low value for a dry air compressor. The working chamber is sealed from its bearings by a combination of lip and labyrinth seals.

Each rotor is supported by one radial and one axial bearing, on the discharge end, and one radial bearing on the suction end of the compressor. The bearings are loaded by a high frequency force, which varies due to the pressure change within the working chamber. Both radial and axial forces, as well as the torque change with a frequency of 4 times the rotational speed. This corresponds to 400Hz and coincides with the number of working cycles that occur within the compressor per unit time.

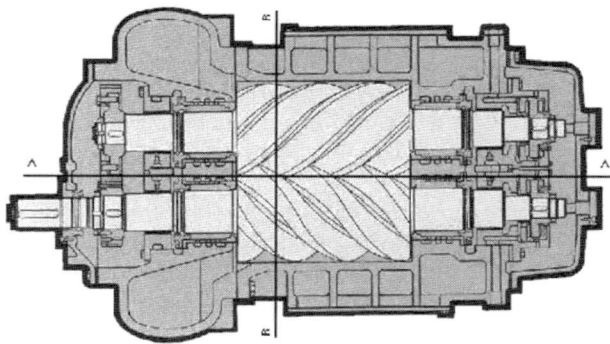

Figure 4-1 Cross section of a dry screw compressor

The compressor takes in air from the atmosphere and discharges it to a receiver at a constant output pressure of 3 bar. Although the pressure rise is moderate, leakage through radial gaps of 150 μm is substantial. In many studies and modelling procedures, volumetric losses are assumed to be a linear function of the cross sectional area and the square root of pressure difference, assuming that the interlobe clearance is kept more or less constant by the synchronising gears. The leakage through the clearances is then proportional to the clearance gap and the length of the leakage line. However, a large clearance gap is needed to prevent contact with the housing caused by rotor deformation due to the pressure and temperature changes within the working chamber. Hence, the only way to reduce leakage is to minimise the length of the sealing line. This can be achieved by careful design of the screw rotor profile. Although minimising leakage is an important means of improving a screw compressor efficiency, it is not the only one. Another is to increase the flow area between the lobes and thereby increase the compressor flow capacity, thereby reducing the relative effect of leakage. Modern profile generation methods take these various effects into account by means of optimisation procedures which lead to enlargement of the male rotor interlobes and reduction in the female rotor lobes. The female rotor lobes are thereby strengthened and their deformation thus reduced.

To demonstrate the improvements possible from rotor profile optimisation, a three dimensional flow analysis has been carried out for two different rotor profiles within the same compressor casing, as shown in Figure 4-2. Both rotors are of the "N" type and rack generated.

Figure 4-2 'N' Rotors, Case-1 upper, Case-2 lower

Case 1 is an older design, similar in shape to SRM "D" rotors. Its features imply that there is a large torque on the female rotor, the sealing line is relatively long and the female lobes are relatively weak.

Case 2, shown on the bottom of Figure 4-2, has rotors optimised for operating on dry air. The female rotor is stronger and the male rotor is weaker. This results in higher delivery, a relatively shorter sealing line and less torque on the female rotor. All these features help to improve screw compressor performance.

The results of these two analyses are presented in the form of velocity distributions in the planes defined by cross-sections A-A and B-B, shown in Figure 4-1.

In the case of this study, the effect of rotor profile changes on compressor integral performance parameters can be predicted fairly accurately with one-dimensional models, even if some of the detailed assumptions made in such analytical models are inaccurate. Hence the integral results obtained from the three-dimensional analysis are compared with those from a one-dimensional model.

4.2.1 Grid Generation for a Dry Screw Compressor

In Case-1, the rotors are mapped with 52 numerical cells along the interlobe on the male rotor and 36 cells along each interlobe on the female rotor in the circumferential direction. This gives 208 and 216 numerical cells respectively in the circumferential direction for the male and female rotors. A total of 6 cells in the radial direction and 97 cells in the axial direction is specified for both rotors. This arrangement results in a numerical mesh with 327090 cells for the entire machine. The cross section for the Case-1 rotors is shown in Figure 4-3. The female rotor is relatively thin and has a large radius on the lobe tip. Therefore, it is more easily mapped than in Case-2 where the tip radius is smaller, as shown in Figure 4-4.

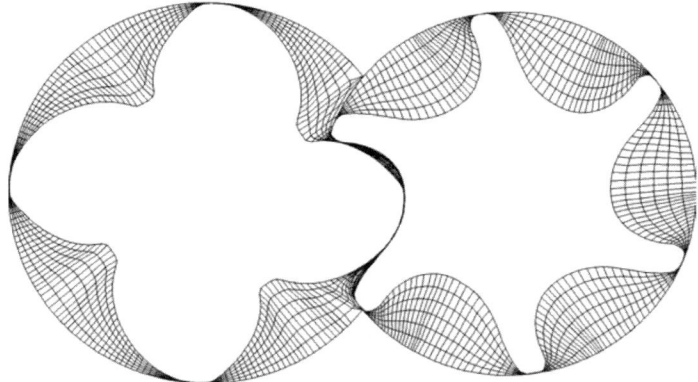

Figure 4-3 Cross section through the numerical mesh for Case-1 rotors

The rotors in Case 2 are mapped with 60 cells along the male rotor lobe and 40 cells along the female lobe, which gives 240 cells along both rotors in the circumferential direction. In the radial direction, the rotors are mapped with 6 cells while 111 cells are selected for mapping along the rotor axis. Thus, the entire working chamber for this compressor has 406570 cells. In this case, different mesh sizes are applied and different criteria are chosen for the boundary adaptation of these rotors. The main adaptation criterion selected for the rotors is the local radius curvature with a grid point ratio of 0.3 to obtain the desired quality of distribution along the rotor boundaries. By this means, the more curved rotors are mapped with only a slight increase in the grid size to obtain a reasonable value of the grid aspect ratio. To obtain a similar grid aspect ratio without adaptation, 85 cells would have been required instead of 60 along one interlobe on the female rotor. This would give 510 cells in the circumferential direction on each of rotors. If the number of cells in the radial direction is also increased to be 8 instead of 6 but the number of cells along axis is kept constant, the entire grid would contain more then a million cells which would, in turn, result in a significantly longer calculation time and an increased requirement for computer memory.

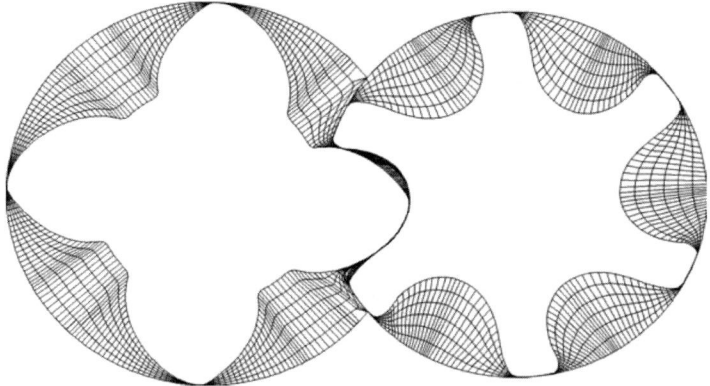

Figure 4-4 Cross section through the numerical mesh for Case-2 rotors

4.2.2 Mathematical Model for a Dry Screw Compressor

The mathematical model used is based on the momentum, energy and mass conservation equations as given in Chapter 2. The equation for space law conservation is calculated in the model in order to obtain cell face velocities caused by the mesh movement. The system of equations is closed by Stoke's, Fourier's and Fick's laws and the equation of state for an ideal gas. This defines all the properties needed for the solution of the governing equations.

4.2.3 Comparison of the Two Different Rotor Profiles

The results obtained for both Case 1 and Case 2 compressors are presented here. To establish the full range of working conditions and to obtain an increase of pressure from 1 to 3 bars between the compressor suction and discharge, 15 time steps were required. A further 25 time steps were then needed to complete the full compressor cycle. Each time step needed about 30 minutes running time on an 800 MHz AMD Athlon processor. The computer memory required was about 400 MB.

88   4 Applications

In Figure 4-5 the velocity vectors in the cross and axial sections are compared. The top diagram is given for Case-1 rotors and the bottom one for Case-2. As may be seen, the Case 2 rotors realised a smoother velocity distribution than the Case 1 rotors. This may have some advantage and could have increased the compressor adiabatic efficiency by reduction in flow drag losses. In both cases, recirculation within the entrapped working chamber occurs as consequence of the drag forces in the air as shown in the figure. On the other hand, different fluid flow patterns can be observed in the suction port. The velocities within the working chambers and the suction and discharge ports are kept relatively low while the flow through the clearance gaps changes rapidly and easily reaches sonic velocity.

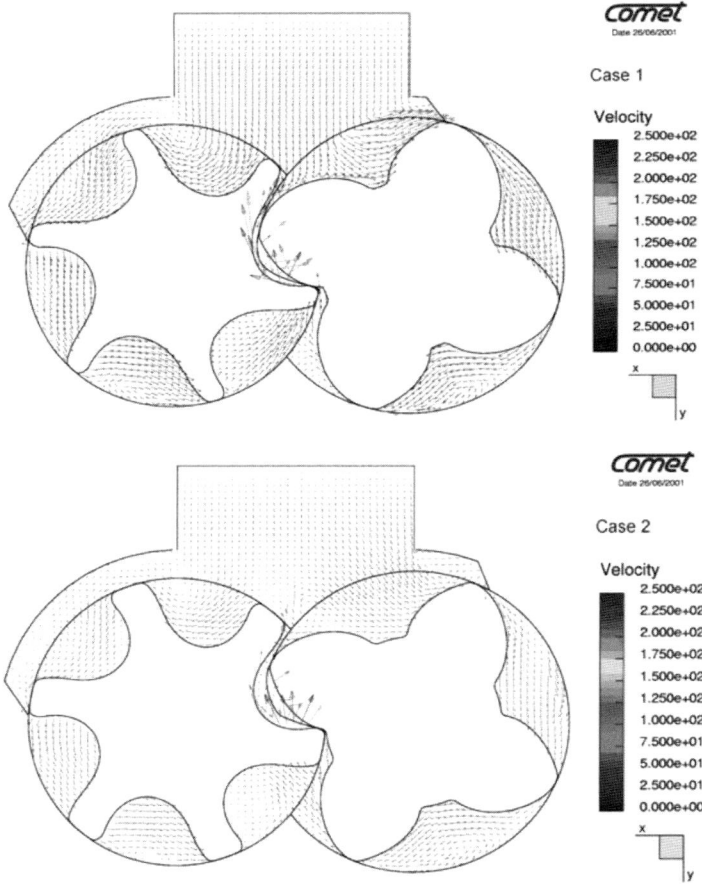

Figure 4-5 Velocity field in the compressor cross section for Case1 and Case2 rotors

4.2 Flow in a Dry Screw Compressor   89

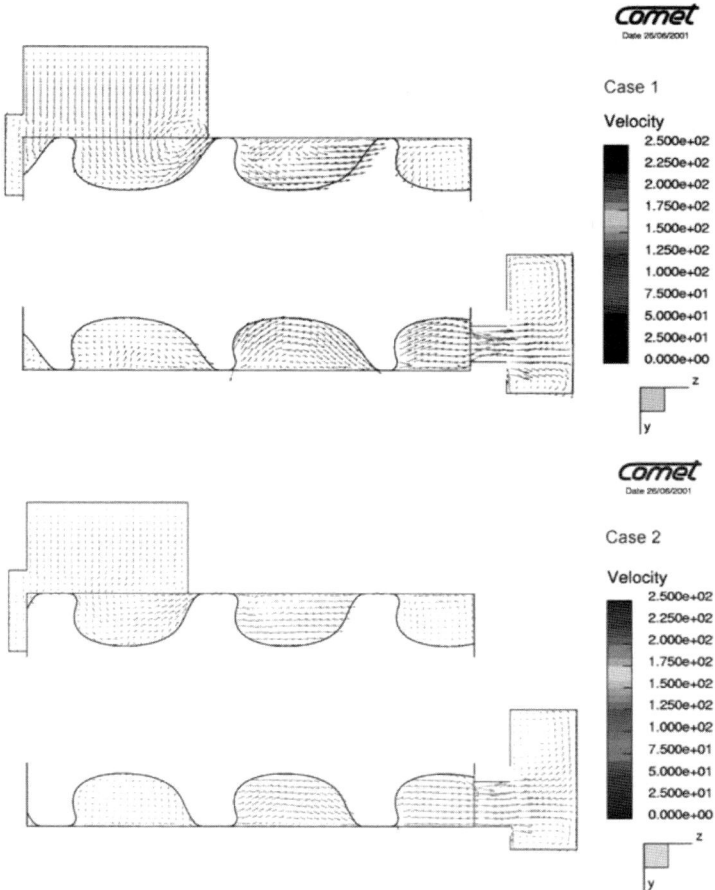

Figure 4-6 Velocity field in the compressor axial section for Case1 and Case2 rotors

These differences are confirmed in the view of the vertical compressor section through the female rotor axis, shown in Figure 4-6. In Case 2, lower velocities are achieved not only in the working chamber but also in the suction and discharge ports. In the suction port, this is significant because of the fluid recirculation which appears at the end of the port. This recirculation causes losses which cannot be recovered later in the compression process. Therefore, many compressors are designed with only an axial port instead of both, radial and axial ports. Such a situation reduces suction dynamic losses caused by recirculation but, on the other hand, increases the velocity in the suction chamber which in turn decreases efficiency. Some of these problems can be avoided only by the design of screw compressor rotors with larger lobes and a bigger swept volume and a shape which allows the suction process to be completed more easily. However, rotor profile design based on existing one-dimensional procedures neglects flow variations in the ports and hence is inferior for this purpose. In such cases, only a full three dimensional approach such as this, will be effective.

Similarly, in the discharge port, velocities are lower for Case 2 then for Case 1. This is an additional advantage of larger flow cross sectional area of rotors, which in turn, gives a larger discharge port through which the same or even a larger amount of gas is delivered at lower speed. Additionally, the longer clearance gaps on the female rotor tip give more resistance to airflow and this reduces the clearance volume losses through this passage.

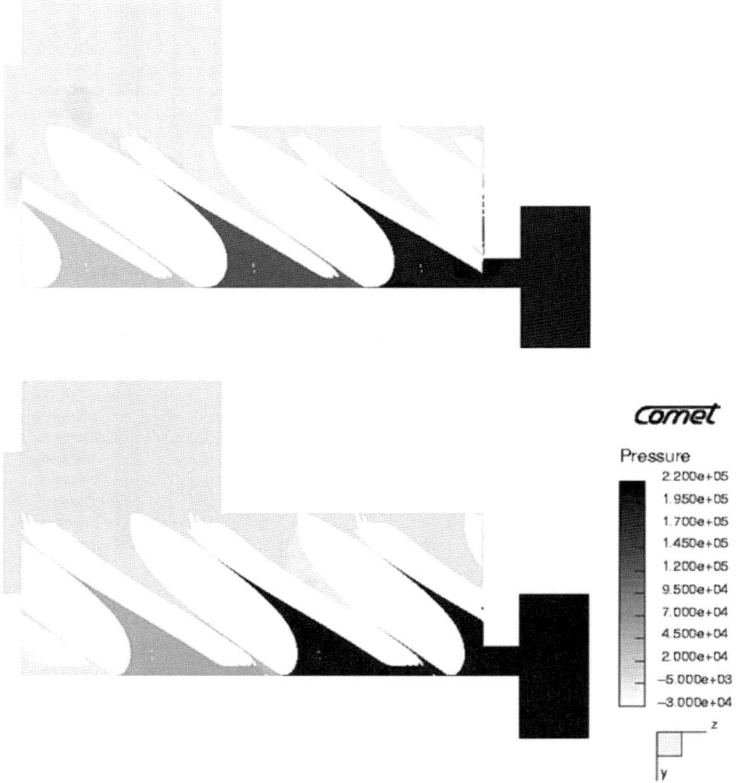

Figure 4-7 Pressure field in the cross section, Case 1 – Top; Case 2 – Bottom

The rise in pressure is similar across the compressor working domains in both cases as presented in Figure 4-7. However, leakages in the second case are smaller due to the shorter sealing line and consequently the pressure rise is slightly steeper. This affects the consumed power, which is lower in Case 2 than in Case 1, despite the higher compressor delivery. These benefits are also visible in Figure 4-9, in which the diagrams of pressure against angle of rotation are presented and compared with each other.

4.2 Flow in a Dry Screw Compressor 91

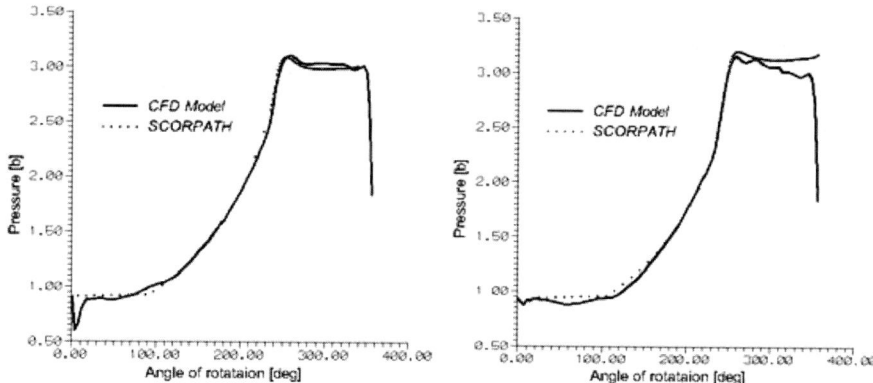

Figure 4-8 Comparison of Pressure-angle diagrams for 1-D and 3-D models

Left - Case 1; Right – Case 2

The 3-D simulations have been validated by comparison with the results obtained from the one-dimensional software, which has been verified and regarded as reliable for compressor design. More about that software can be found in *Stosic et al* (2005). Results of the comparison are shown in Figure 4-8.

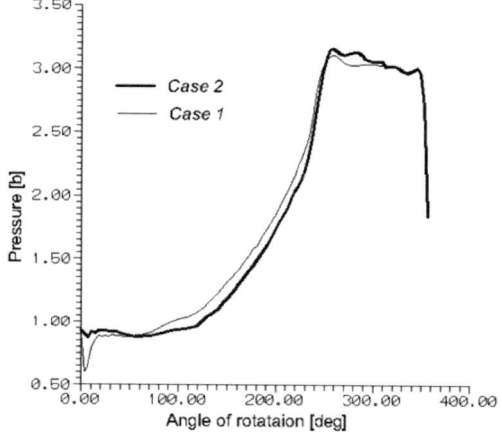

Figure 4-9 Comparison of Pressure-angle diagrams for both cases

Agreement of the presented results is good for the compression within the entrapped chamber represented by the middle part of the p-α diagram. However, some differences are visible in the suction and discharge regions. A higher pressure drop is obtained with the 3-D model than with the 1-D model for both rotor cases. This may be due to the inability of a 1-D model to predict dynamic flow losses in both the suction and discharge ports caused by substantial recirculation,

92   4 Applications

as shown in Figure 4-6. This opens many questions about how the shape and position of the inlet and outlet ports can be altered in order to improve screw compressor design. Thus, the mathematical model established in this book and the grid generating tool developed for that purpose can be used to explore this further.

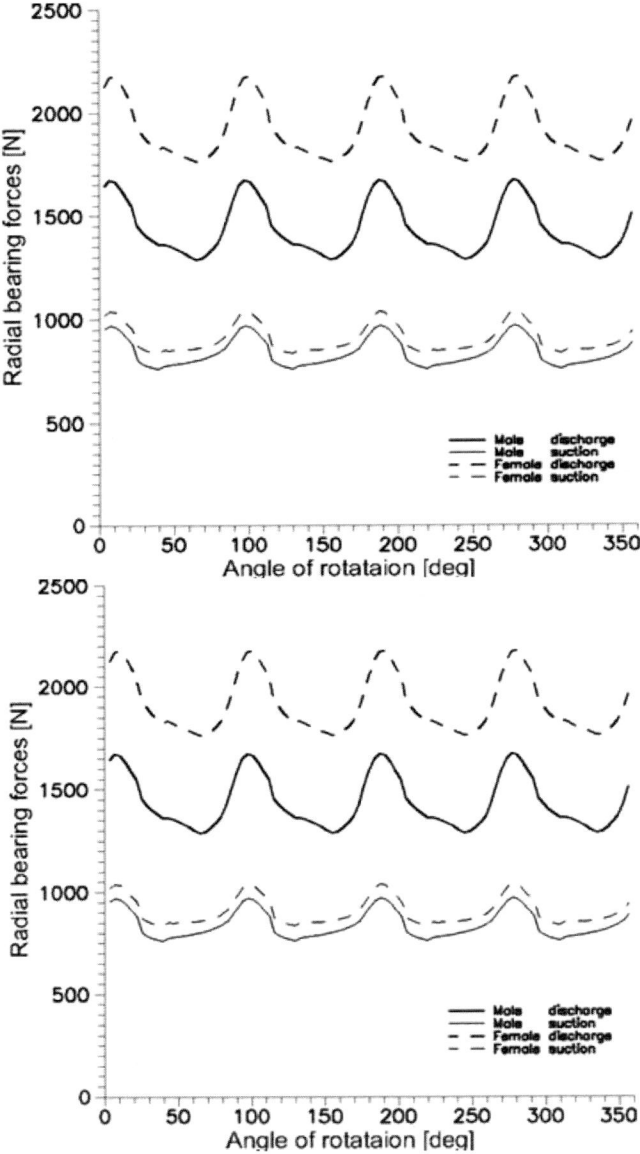

Figure 4-10 Radial bearing forces, Case 1 –top, Case 2 –bottom

4.2 Flow in a Dry Screw Compressor   93

Forces on the bearings are a consequence of the pressure differences within the screw compressor working chambers. Radial bearing forces on both the suction and discharge sides are presented in Figure 4-10.

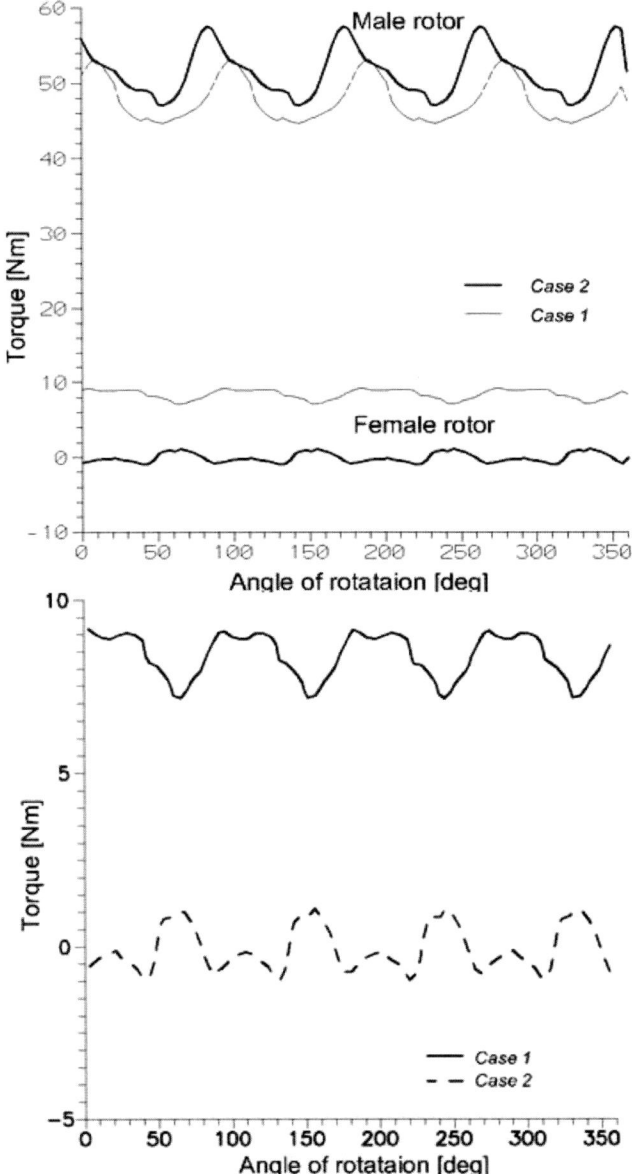

Figure 4-11 Comparison of the compressor torque in Case 1 and Case 2
Top – Both rotors; Bottom – Female rotors

The average values are practically the same in both cases. However, the distribution of forces upon the suction and discharge sides differs for the two cases analysed. The forces on both the male and the female radial bearings are lower in Case 2 than in Case 1, while the bearings on the discharge side are loaded with almost equal forces in both cases. Although the radial forces are similar, the torque is significantly different for the two cases. The torque is a function of the compressor suction and discharge pressures, which, according to Figure 4-9 are almost the same. However, the profile shape is significantly different for these two cases and consequently the distribution of torque on the male and female rotors is not the same. In Case 2 the torque on the male rotor is higher than in Case 1 but the torque transmitted to the female rotor is significantly lower, Figure 4-11. The bottom diagram of that figure shows that the torque on the female rotor is reduced from 10 (Nm) in Case 1 to practically zero in Case 2, while at the same time the female rotor is thicker. The increased thickness gives many advantages to the rotors in Case 2. Among others, the most important features are: lower female rotor deformation caused by the pressure, a greater flow cross sectional area and a shorter sealing line.

All the features listed above, confirmed by the diagrams, obtained from the 3-D CFD calculations, show that the modern rotors in Case 2 have significant advantages over the more traditional rotors of Case 1. This is also confirmed by comparison of the integral parameters given in Table 4-1. Here it can be seen that the specific power of the compressor with Case 2 rotors is 12% lower then in the other case. This indicates that both the input power is lower and the delivered flow rate is greater. This is confirmed by a 12% higher volumetric efficiency η_v and a 10% higher thermal efficiency η_i for the compressor with the Case 2 rotor profile.

Table 4-1 Comparison of the integral parameters for the two cases

	$\dot{V}$ (m³/min)	P (kW)	P_{spec}(kW/m³min)	ηv (%)	ηι (%)
Case 1	17.3	56.8	3.28	74.4	61.0
Case 2	19.6	55.9	2.84	83.9	67.5

4.3 Flow in an Oil Injected Screw Compressor

Figure 4-12 shows an oil-injected screw compressor, designed and built at City University. The cross section in Figure 4-13 shows how the two rotors are supported by six bearings. Two of these are thrust bearings on the discharge side on the far right of the figure, while the others are cylindrical roller bearings. Oil is supplied through an injection port in order to seal, cool and lubricate the rotors. The same oil is supplied to the bearings, in order to lubricate them. Therefore, no seals are required within the compressor. Only the driving shaft requires an external shaft seal to protect the compressor from the surroundings, as presented on the far left of this figure. Three cross sectional planes are indicated in Figure 4-13. Two of them are normal to the rotor axes indicated by A-A and B-B. The first one crosses the suction port, rotors and the oil injection port while the other is closer to the discharge port. The remaining section is parallel to the rotor axes lying between the rotors. All results in this example are presented for these three cross sections.

Figure 4-12 Oil injected screw compressor with 'N' rotors.

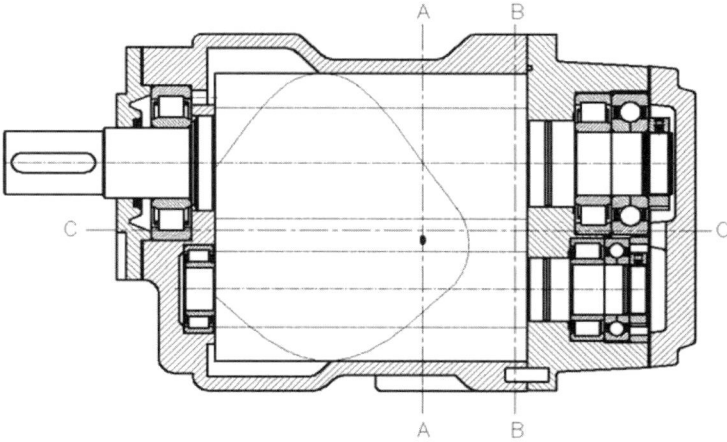

Figure 4-13 Section of the analysed oil injected compressor

The Rotor profiles are of the 'N' type with a 5/6 lobe configuration. The rotor outer diameters are 128 and 101 mm for the male and female rotors respectively, and their centre lines are 90 mm apart. The rotor length to diameter ratio is 1.66. Both, a drawing and photograph of the rotors are presented in Figure 4-14.

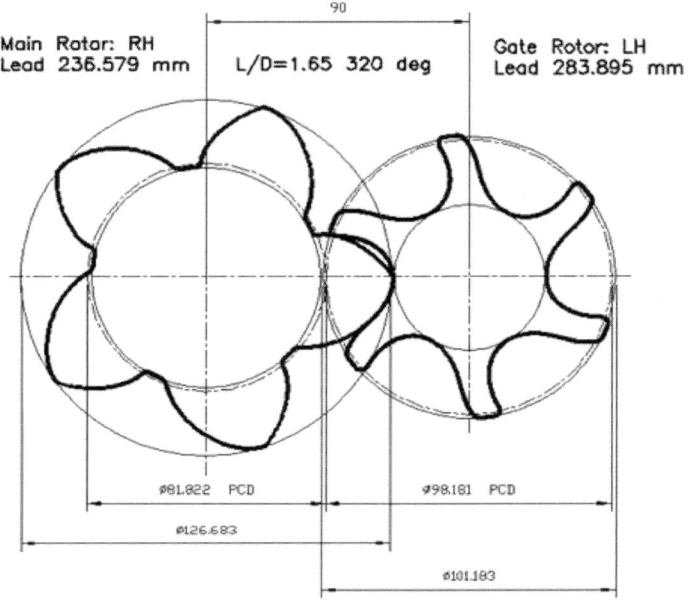

Figure 4-14 Drawing and photograph of 5/6 male and female 'N' rotors

4.3.1 Grid Generation for an Oil-Flooded Compressor

The male and female rotors have 40 numerical cells along each interlobe in the circumferential direction, 6 cells in the radial direction and 112 in the axial direction. These form a total number of 444,830 cells for both rotors and the housing. To avoid the need to increase the number of grid points, if a more precise calculation is required, an adaptation method has been applied to the boundary definition.

The number of time changes was 25 for one interlobe cycle in this case. The total number of time steps needed for one full rotation of the male rotor is then 125. The number of cells in the rotors was kept the same for each time step. To achieve this, a special grid moving procedure was developed in which the time step was determined by the compressor speed, as explained in Chapter 4. The numerical grid for the initial time step is presented in Figure 4-15.

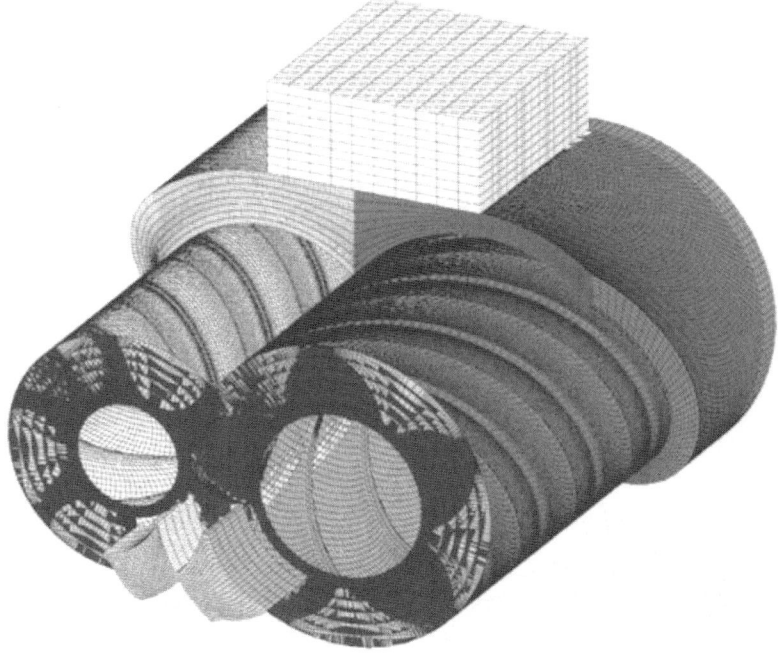

Figure 4-15 Numerical grid for oil injected screw compressor with 444,830 cells

4.3.2 Mathematical Model for an Oil-Flooded Compressor

The mathematical model consists of the momentum, energy, mass and space equations, as described in section 2.2, but an additional equation for the scalar property of oil concentration was added to enable the influence of oil on the entire compressor performance to be calculated. The constitutive relations are the same as in the previous example. The oil is treated in the model as a 'passive' species, which does not mix with the background fluid - air. Its influence on the air is accounted for through the energy and mass sources which are added to or subtracted from the appropriate equation of the main flow model. In this case, the momentum equation is affected by drag forces as described earlier.

To establish the full range of working conditions and starting from a suction pressure of 1 bar to obtain an increase in pressure of 6, 7, 8 and 9 bars at discharge, a numerical mesh of nearly 450,000 cells was used. For each case only 25 time steps were required to obtain the required working conditions, followed by a further 25 time steps to complete a full compressor cycle. Each time step needed about 30 minutes running time on an 800 MHz AMD Athlon processor. The computer memory required was about 450 MB.

4.3.3 Comparison of the Numerical and Experimental results for an Oil-Flooded Compressor

In the absence of velocity field measurements in the compressor chamber, an experimentally obtained pressure history within the compressor cycle and the measured air flow and compressor power served as a valuable basis to validate the results of the CFD calculation. To obtain these values, the 5/6 oil flooded compressor, already described, was tested on a rig installed in the compressor laboratory at City University London, Figure 4-16.

Figure 4-16 Oil-Injected air screw compressor 5/6-128mm (a=90mm) in the test bed

4.3 Flow in an Oil Injected Screw Compressor

The test rig meets all Pneurop/Cagi requirements for screw compressor acceptance tests. The compressor was tested according to ISO 1706 and its delivery flow was measured following BS 5600.

The pressures were measured with high quality pressure transducers, with readings taken at the inlet to the compressor, discharge from the compressor and in the separator.

The temperatures were measured by FeCo thermocouples at the inlet to and discharge from the compressor and after the oil separator. Measurements of temperature were also taken of both, the oil and the cooling water at the inlet end of the oil cooler. The oil flow rate was calculated from the cooler and compressor energy and mass balances.

Torque was measured by a laboratory type torque meter transducer IML TRP-500 connected between the engine and the compressor driving shaft. The compressor was driven by a diesel engine prime mover of 100 kW maximum output, which could operate at variable speed. The compressor speed was measured by a frequency meter and the signal was transferred to a data logger after converting to current.

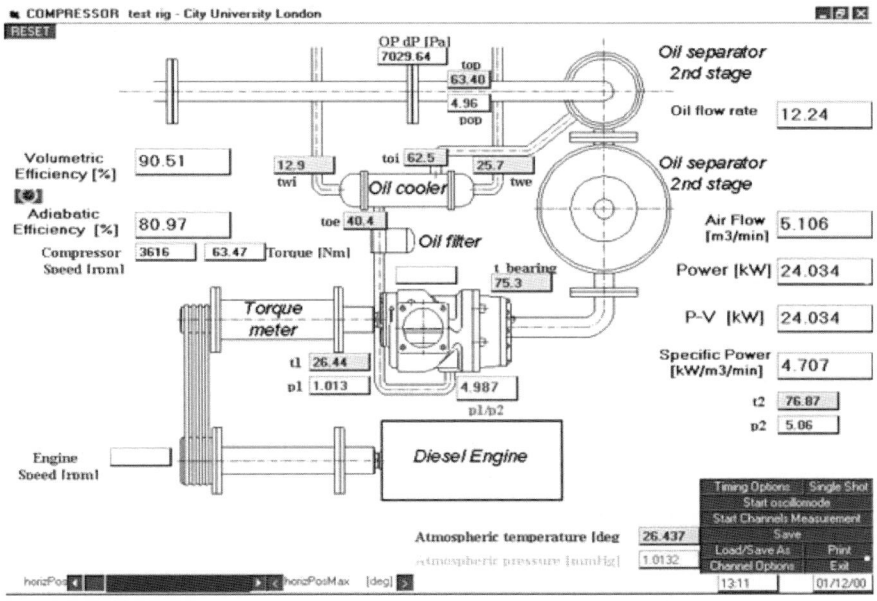

Figure 4-17 Computer screen of compressor test rig measuring program

The compressor flow was measured by an orifice plate according to BS 5600 with the differential pressure measured by a pressure transducer PDCR 120/35WL over an operating range of 0-200 kPa.

The measured values of all relevant pulsating quantities were used to obtain details of the thermodynamic cycle. Of these, the pressure in the trapped volume

100   4 Applications

was the most significant since it was required to plot the machine p-V diagram. Accordingly, a method was developed to construct an entire p-V diagram from the recording of pressure changes at only 4 discrete points in the machine casing.

Endevco piezoresistive transducers E8180B were used to measure the instantaneous values of the absolute pressure in the compressor. Each transducer recorded the pressure in one interlobe space. Starting from the suction end, 4 transducers were positioned in the compressor casing to record the changes in each consecutive interlobe space. When plotted in sequence they gave a pressure-time diagram for the whole compressor working cycle.

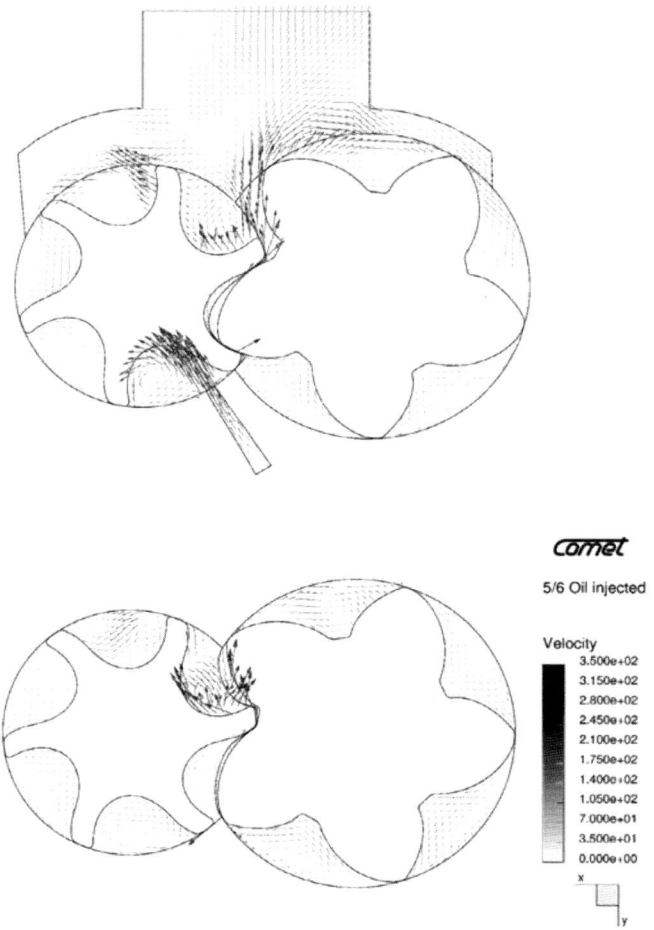

Figure 4-18 Velocity vectors in the two compressor cross sections
Top – cross section A-A through the suction port, Bottom – cross section B-B

All measured values were automatically logged and transferred to a PC through a high-speed InstruNet data logger. The data acquisition system enabled high speed measurements to be made at frequencies of more then 2 kHz. An acquisition and measuring program for the PC was written for this in Visual Basic that permitted online measurement and calculation of the compressor working parameters. A computer screen record of this measuring program is given in Figure 4-17.

In Figure 4-18 the velocity vectors in two cross sections are presented. One of these is through the inlet port and oil injection pipe and the other is close to discharge. Figure 4-19 shows the velocities in the vertical section through the compressor. High velocity values in the gaps, both between the rotors and their housing and between the two rotors, are generated by the sharp pressure gradients through the clearances. These are clearly distinguished from the velocities in the interlobe regions where the fluid flows relatively slowly. The fluid flow is caused there only by movement of the numerical mesh, which is generated in a manner to follow the movement of the rotors in time. The top diagram shows the cross section through both the suction port and oil injection openings. Recirculation in the suction port is substantial and seems to be high because of the position of the oil injection hole. If the oil injection had been positioned further downstream, the recirculation would have been reduced. The bottom diagram, which shows a cross section close to the discharge port, indicates that more recirculation is present in the lobes with lower pressures, as is visible in the top of the diagram. The velocities in the high pressure regions are smoothed to relatively low values, to some extent similar to the wall velocities.

The velocity field in the axial section C-C, which crosses both rotors along the rotor bore cusp, is shown in Figure 4-19.

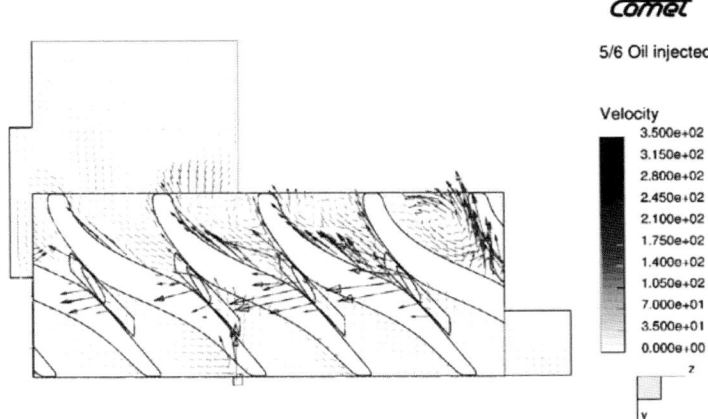

Figure 4-19 Velocity vectors in the compressor axial section C-C

Smoothing of the velocities is visible in the high pressure regions at the right end of the figure. In the upper portions of the compressor, where both, low pressures and low pressure gradients occur, flow patterns are more curved, thus indicating flow swirls. There is also recirculation in the far end of the suction port while, at the same time, the flow through the axial part of the port is more intensive.

102 4 Applications

The oil distribution and pressure field in the cross section A-A are shown on the top and bottom diagrams of Figure 4-20 respectively. As noted earlier, some fluid recirculates from the working chamber to the suction port through the compressor clearances. Figure 4-20 indicates that together with air, the oil escapes from the pressurised working chamber to the suction port through the rotor-to-rotor leakage paths. The presence of oil in the suction port was also observed visually during tests on this compressor. However, no measurements were made of it.

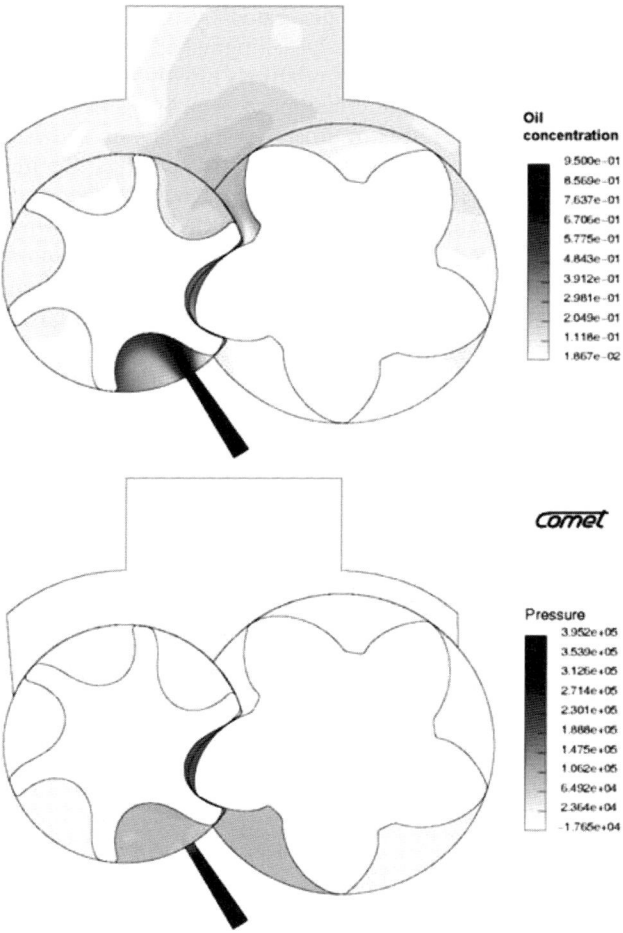

Figure 4-20 Cross section through the inlet port and oil injection port A-A
Top – mass concentration of oil, Bottom - Pressure distribution

Some limited results of an experimental investigation on oil distribution within a screw compressor are published by *Xing et al* (2001). In that case, the oil flow was observed by making the compressor casing from a transparent material. Although the authors do not have a complete record of their results, it appears from what they published that the oil flow patterns obtained from the 3-D calculations are similar to those obtained in their experiments. The presence of hot oil in the suction port, although beneficial for the lubrication of the rotors, increases the gas temperature before the working chamber is closed. This reduces the trapped mass and hence the compressor capacity and is another of the effects which are not modelled by one-dimensional models of screw compressor processes.

Figure 4-21 shows the pressure distribution within the compressor with a male rotor speed of 5000 rpm. This figure indicates that the pressure within the each working chamber is almost uniform and that it can be regarded as such for almost all calculations and comparisons. Due to that, the results obtained from the 3-D calculations may be compared with those obtained from measurements.

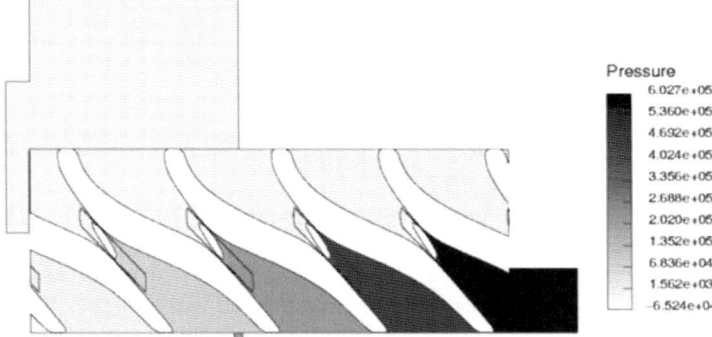

Figure 4-21 Axial section between two rotors - Pressure distribution

The change in pressure within the working chamber is shown in Figure 4-22 as a function of the male rotor shaft angle. Here the pressure-shaft angle diagrams are compared with results from the compressor tests. The results shown are for discharge pressures of 6, 7, 8 and 9 bar absolute at a shaft speed of 5000 rpm. In all cases, the inlet pressure was 1 bar. The agreement between the predicted and measured values is reasonable, especially during the compression process. Some differences are recorded in the suction and discharge regions. Those in the suction region are probably the consequence of the flow fluctuations visible in Figure 4-19, which shows that the flow during suction and at the very beginning of the compression is not so damped. On the other hand, the piezoresistive transducers used for the measurement of pressure are subjected to a higher error at lower pressure differences, which are close to zero in these areas. The differences recorded at the high pressure end, during the discharge process, are probably generated because of the inability to capture real geometry accurately. The calculated discharge port was simplified from the real one. It was also mapped with a relatively low number of cells. The influence of the mesh size on the calculation accuracy is analysed in more detail in section 4.3.5.

104 4 Applications

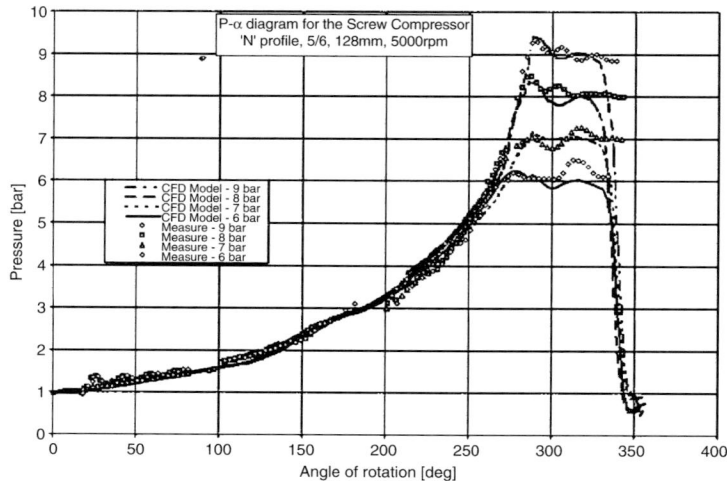

Figure 4-22 Pressure-shaft angle diagram; comparison of CFD calculations and measurements

The compressor radial bearing forces are presented in Figure 4-23. There, the force on the female rotor radial discharge bearing and the force on the male rotor radial suction bearing are given. The reason why these two radial bearings were selected for the diagram is that each of them is loaded with a higher force then its counterpart on the other rotor. It can be seen that both the mean value and amplitude of the forces rise with the discharge pressure. More frequent dynamic load with higher amplitude puts a greater demand on the bearings. In compressor design, these have to be selected carefully to withstand such loads.

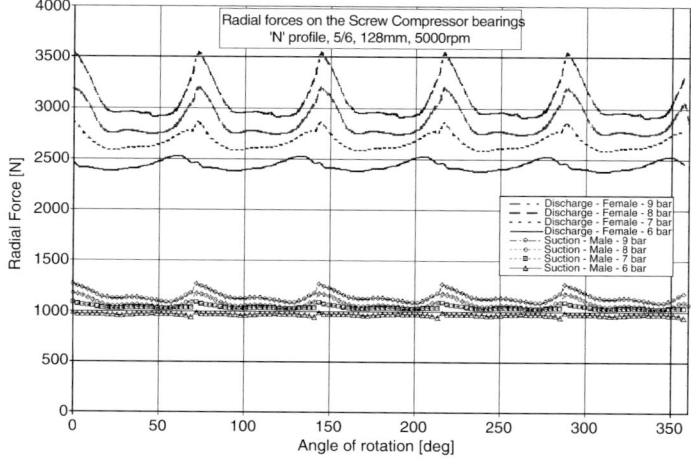

Figure 4-23 Radial bearing forces acting on supporting bearings

4.3 Flow in an Oil Injected Screw Compressor

Diagrams of the torque on the male and female rotors versus the shaft angle, are shown in Figure 4-24. Changes in torque have the same frequency as those of the radial force or any other parameter of screw compressor processes that correspond to the product of the shaft speed and the number of lobes on the compressor male rotor. It is very important for a compressor, which works in the oil flooded operational mode, to have as low as possible torque on the female rotor which is usually driven by the male rotor by direct contact. The rotor profile of this compressor is such that the female rotor operates with a very small positive torque.

The results obtained from the CFD simulation model are later used to calculate such compressor integral parameters as the delivered flow rate and consumed power input. From them, the specific power and efficiencies are obtained by a procedure described in section 2.5. The estimated and measured values of the compressor delivery and input power are presented and compared in Figure 4-25 for all selected discharge pressures.

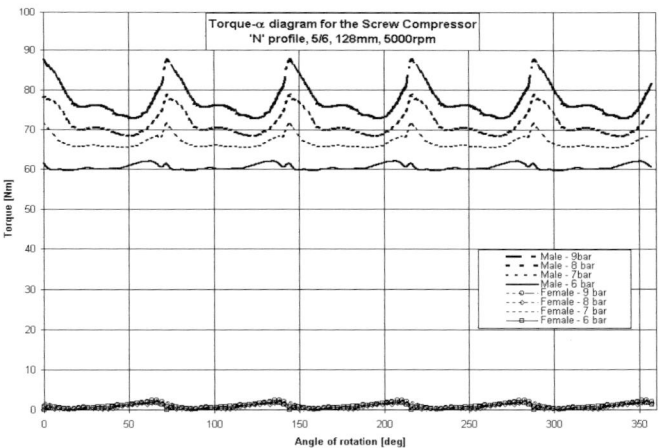

Figure 4-24 Torque on the male and female rotors

The maximum difference between the estimated and measured power input is 8%, while the values of the compressor flow differ by up to 10%. The difference is larger for low discharge pressures where the calculations gave lower flow rates and higher power consumption than were measured. It implies that some effects of the compressor flow are not estimated well enough. The results presented for this compressor were obtained by calculation assuming the flow to be laminar. In the following section, a k-ε model of turbulence is incorporated in the flow calculation and the results are compared with the laminar flow calculations.

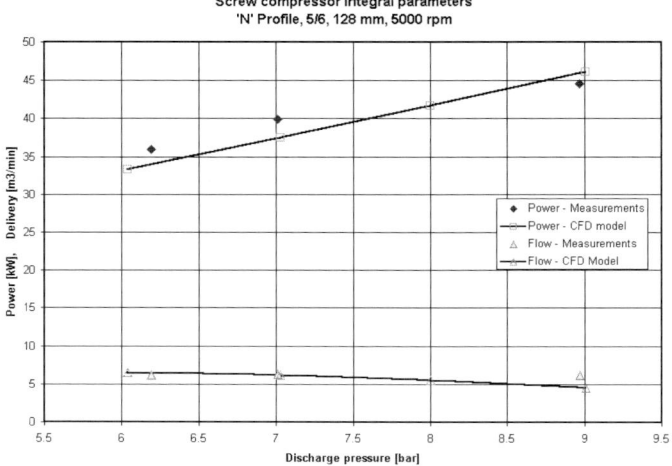

Figure 4-25 Comparison of the integral parameters at 5000 rpm shaft speed

4.3.4 Influence of Turbulence on Screw Compressor Flow

A standard k-ε model of turbulence is applied here and the results of the laminar and turbulent flow calculations obtained by these models within the same screw compressor are compared. Turbulence is implemented through the two additional governing equations of kinetic energy of turbulence and its dissipation, as explained in 2.2.5. These two equations are solved separately and the k and ε values thus derived are used to balance the momentum and pressure equations through turbulence viscosity in the next iteration step. The next step approach is applied for two reasons. Firstly, profiles of turbulent kinetic energy and its dissipation contain more peaks than the main velocity profile and these are difficult to capture. Secondly, non-physical negative values of k and ε can possibly appear in some cases which lead to numerical instability. An additional problem in the implementation of the turbulence model to screw compressor calculations is the very high value of the ratio of the main compressor chamber dimensions to the clearances. In order to maintain a block structured configuration of the numerical mesh, the number of cells in the main domain and in the clearances must be the same. This means that the numerical cells within a compressor change in size by about the same ratio. The kinetic energy of turbulence and especially its dissipation are both 'stiff' in such situations, which easily leads to calculation instability and excessive rise in dissipation rate. To explore such a situation, simple approximate calculations are applied to estimate the dissipation level and the scale of turbulence, which are similar to calculations performed by *Lumley* (1999).

Turbulence is a phenomenon in which a part of mechanical energy of the main fluid flow is transformed to internal energy. It consists of three phases, namely production, transport and dissipation. In the production phase the kinetic energy of the main flow is transferred to the largest eddies that possibly can appear in the flow. Then the energy is passed on to eddies of progressively smaller scale until the smallest value is reached. This constitutes the transport phase of turbulence. The energy passed to the smallest eddies is transformed to heat through the dissipation phase. The amount of energy of turbulence dissipated per unit mass is called the dissipation ε. It is proportional to the energy of the turbulence and the natural time scale of the most energetic eddies. It can be expressed as:

$$\varepsilon = u^2 \frac{u}{l} = \frac{u^3}{l} \sim \frac{k^{\frac{3}{2}}}{l}, \tag{4.1}$$

where u is the velocity, l is the length scale of the most energetic eddies and k is the kinetic energy of turbulence. The smallest scale at which the energy is transferred to heat is called the Kolmogorov microscale,

$$\eta = \sqrt[4]{\frac{v^3}{\varepsilon}} \tag{4.2}$$

where v is the kinematic viscosity. According to *Lumley* (1999), it is an experimental fact that the most energetic eddies, responsible for most of the turbulent transport, are about 1/6 of the size of the largest eddy. For a screw compressor this will be 1/6 of an interlobe or 1/6 of the clearance in the gaps. Bearing all these factors in mind, the following estimate is valid for the compressor under consideration here.

In the interlobe: The size of an interlobe is obtained as the difference between the outer and inner rotor radii, Δ=*24.1 mm*. This can be regarded as the size of the largest eddy which can fit in the interlobe. The size of the most energetic eddy will thus be $l \approx 4 \, mm$. The fluid velocity in that area is of the same order of magnitude as the rotor velocity. The male rotor rotates at 5000 rpm which gives *33.2 m/s* tip speed and *20 m/s* root speed. Assume that the turbulent velocity is twice the mean value of these two velocities, i.e. $u \approx 50$ m/s. The kinematic viscosity of air at 20°C suction temperature is $v = 1.51 \times 10^{-5} \, m^2/s$. This gives a dissipation rate of $\varepsilon \approx 3 \times 10^7 \, W/kg = 30 \, MW/kg$, and the length scale of the smallest eddy $\eta \approx 2 \times 10^{-6} \, m = 2 \, \mu m$. The value of 30 *MW/kg* is enormous but the mass of air in the interlobe is only about 2 *g* and it remains there for very short time, in this case only about 2 *ms*.

In the clearance: Assume the mean clearance size in this compressor is δ=*100* μm. The size of the most energetic eddy in this case is $l \approx 15 \, \mu m$. The velocity in the clearances rises to the speed of sound, which for a suction temperature of 20°C has the value of *341 m/s*. For a discharge temperature of 150°C, the speed of sound is *422 m/s*. If the density at discharge has increased six times from the suction value, the kinematic viscosity becomes $v \approx 5 \times 10^{-6} \, m^2/s$. In that case, the dissipation rate increases to the enormous value of $\varepsilon \approx 4 \times 10^{12} \, W/kg$ while the Kolmogorov microscale drops to a fraction of a micrometer, say $0.1 \, \mu m$. One should bear in mind that the leakage flow is less then 10% of the main flow which means that the mass in the clearances is one order of magnitude lower than that of the main flow. In addition, the time for which the fluid remains in the clearance gaps is only a fraction of the compressor cycle. Due to that, the energy wasted by turbulence in clearances is not large.

The Reynolds number of turbulence in the main domain and in the clearances can be calculated from the above values using:

$$\frac{\eta}{l} = \left(\frac{v^3}{\varepsilon l^4}\right)^{1/4} = \left(\frac{v^3}{u^3 l^3}\right)^{1/4} = R_l^{-\frac{3}{4}}. \tag{4.3}$$

In the main domain, the Reynolds number of turbulence is $R_l \approx 2.5 \times 10^4$ and in the clearances $R_l \approx 8 \times 10^2$. Assuming similarity of molecular and turbulent transport of momentum for gases, one can state that the effective viscosity is proportional to the velocity scale and the length scale of the process responsible for transport. Therefore, the measure of molecular transport is v and the measure of turbulent transport is $v_T \approx ul$. Then,

$$\frac{v_T}{v} = \frac{ul}{v} = R_l. \tag{4.4}$$

Using the previous equation, values for turbulent or eddy viscosity are obtained in the main domain and in the clearances. These are $v_T = 3.8 \times 10^{-1} \, m^2/s$ in the main domain and $v_T = 1.2 \times 10^{-2} \, m^2/s$ in the clearances. These are far larger than the molecular viscosity.

One more measure can be introduced here. This is the relative distance for which the turbulence will carry a property. That property can be either momentum or the fluid itself. The relative distance can be estimated from:

$$L \approx \sqrt{\frac{2}{3} v_T \tau}, \tag{4.5}$$

where τ is the time for which the turbulence remains in a particular place. Therefore, in the interlobe, turbulence will carry a property for $\frac{L}{\Delta} \approx \sqrt{\frac{2}{3} u \frac{l}{\Delta^2} \tau_\Delta} \approx 0.68$, while in the clearances $\frac{L}{\delta} \approx \sqrt{\frac{2}{3} u \frac{l}{\delta^2} \tau_\delta} \approx 0.63$. From these two values, it may be concluded that turbulence plays some role in screw compressor processes. It is more important in the suction domain and interlobes exposed to the suction where turbulent viscosity can be four orders of magnitude higher than its molecular equivalent and where a property is transported by turbulence for about two thirds of the available interlobe space. Turbulence dies out in the compression chambers as the pressure increases and the velocity decreases and has no significant influence in the discharge domain. In the clearances, turbulent transport exists, but it lasts for an extremely short time and produces no excessive heat. However, the fluid is transported by turbulence for approximately two thirds of the available space in the clearances.

The values calculated above give an insight into the problems associated with turbulence. The turbulence effects here are estimated by a turbulence model based on the Reynolds averaged Navier-Stokes equations. Both, the kinetic energy of turbulence and its dissipation rate are high in certain regions of the compressor and remain there for an extremely short time. Their exact calculation requires a numerical mesh with a large number of numerical cells and a very short time step. As it was not practical to perform such calculations, the standard k-ε model is used here to show the ability of the model and to give values of the local and integral parameters of the calculated flow. These are compared with the results of calculations assuming laminar flow. The development of a turbulence model suitable for positive displacement machines is recommended for the future.

In order to compare the significance of turbulent and laminar flow assumptions on compressor performance predictions, without including secondary influences, a further analysis was made on the compressor referred to earlier, in which the presence of oil was excluded. The numerical mesh taken for this purpose contains 236,780 numerical cells. The discharge pressure was kept at 3 bar. The low discharge pressure was preferred because of the larger difference in flow parameters between laminar flow calculations and measurements noticed under these conditions. It seems that turbulence has more significant influence under these circumstances. The compressor speed was 5000 rpm.

The pressure distribution and velocity field for both the laminar and turbulent flow cases are compared in Figure 4-26. The laminar results on the top diagram and the turbulent ones on the bottom diagram are not significantly different. Higher velocities in the interlobes at low pressure are recorded with laminar flow calculations while the pressure fields seem to be very similar.

The pressure rise in the compressor is compared in more detail in the pressure-angle diagram,

Figure 4-27. Only a small difference is recorded between the two flows where a slightly higher pressure was obtained with the turbulent flow model. This gives the impression that turbulence has some influence on the calculation in the leakage path area. Turbulence reduces the leakage flow, which results in more delivery and raises pressure in the working chamber. However, the difference in power obtained by this means was very small, being only 0.5%, as shown in Table 4-2.

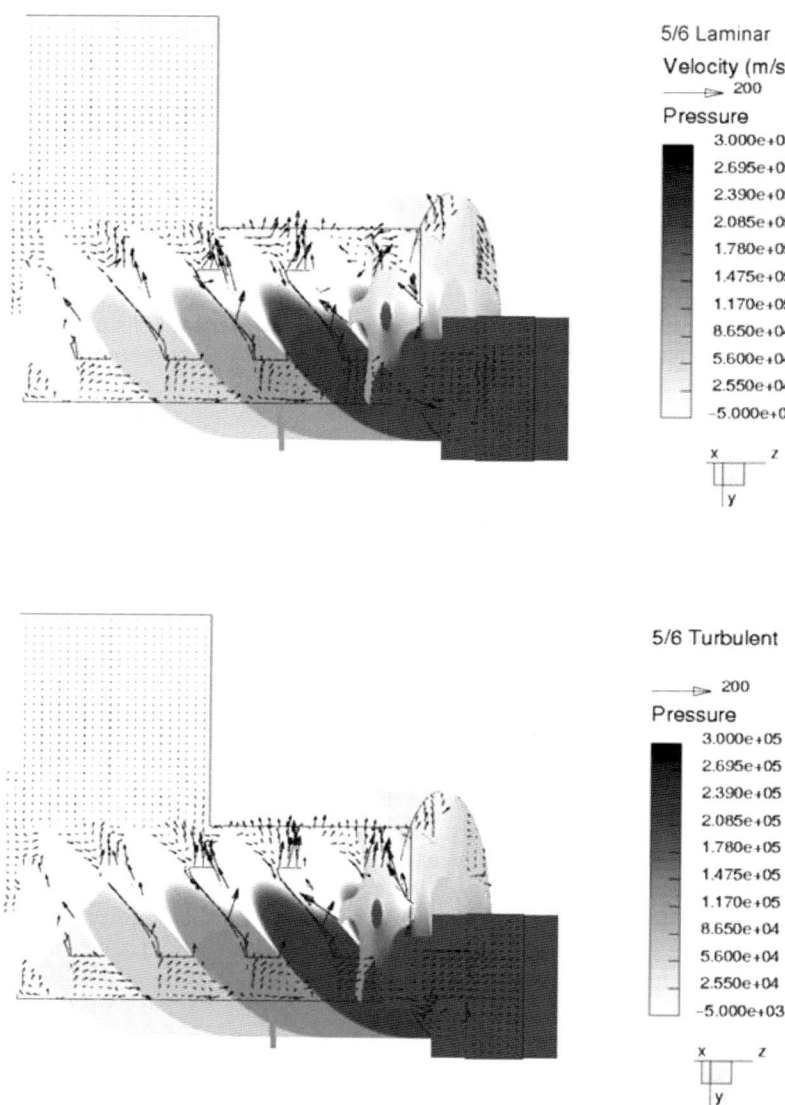

Figure 4-26 Comparison of pressure field and velocity vectors across the screw compressor
Top – laminar model; Bottom – turbulent model

4.3 Flow in an Oil Injected Screw Compressor 111

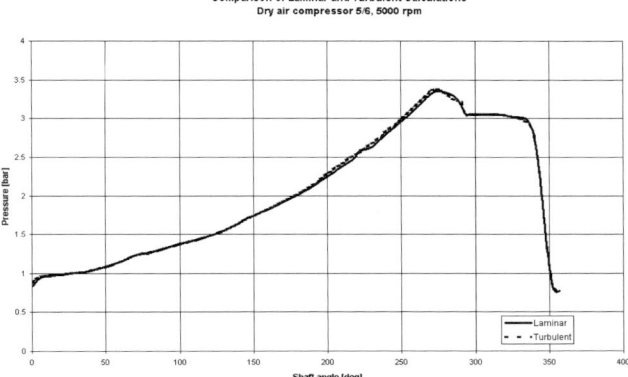

Figure 4-27 Comparison of pressure change for turbulent and laminar flow calculations

The difference in the compressor flow obtained from laminar and turbulent calculations is presented in Figure 4-28. The mass flows at suction and discharge are given as functions of the shaft angle. On average, 4% higher flow is calculated with the turbulent model. The difference was greater at the discharge end of the compressor, both in the mean value and in the amplitude. This agrees with the results obtained from the approximate calculations where turbulent transport through clearances is significant. The difference in flow obtained at the suction end is, on average, less than 3%. This shows that a compressor with a large suction opening has no significant dynamical losses, although turbulence exists in the compressor low pressure domains. It is expected that the difference between the laminar and turbulent flow calculations will be smaller for higher discharge pressures and lower compressor speeds.

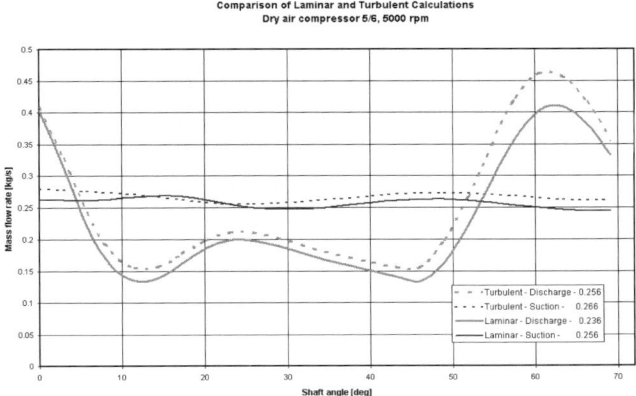

Figure 4-28 Comparison of fluid flow at inlet and exit of screw compressor

The integral parameters obtained from both the laminar and turbulent numerical models are presented in Table 4-2. According to these results, it can be concluded that turbulence has some influence on the screw compressor. Its effect is greater at lower pressure ratios and low compressor speeds.

Table 4-2 Integral parameters calculated with laminar and turbulent model

	Flow rate (m3/mim)	Power (kW)	Specific power (kW/m3/min)	Out temp (K)
Laminar	2.749	15.57	5.684	402.3
Turbulent	2.861	15.66	5.473	402.5

More detailed insights into the results obtained from the k-ε model of turbulence can be found in the following four figures; Figure 4-29 shows the kinetic energy of turbulence. The dissipation rate is presented in Figure 4-30, the turbulent viscosity in Figure 4-31 and the dimensionless distance from wall y+ is given in Figure 4-32.

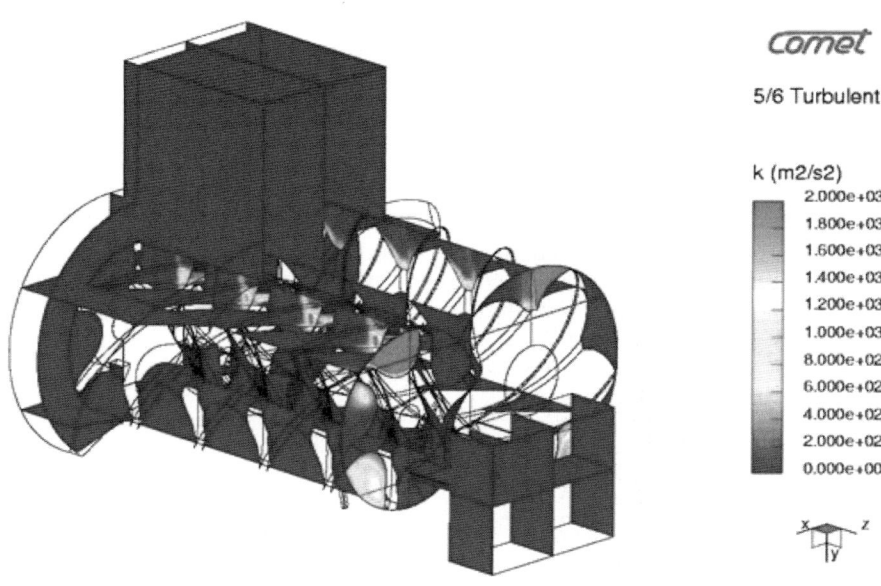

Figure 4-29 Kinetic energy of turbulence within the screw compressor

4.3 Flow in an Oil Injected Screw Compressor   113

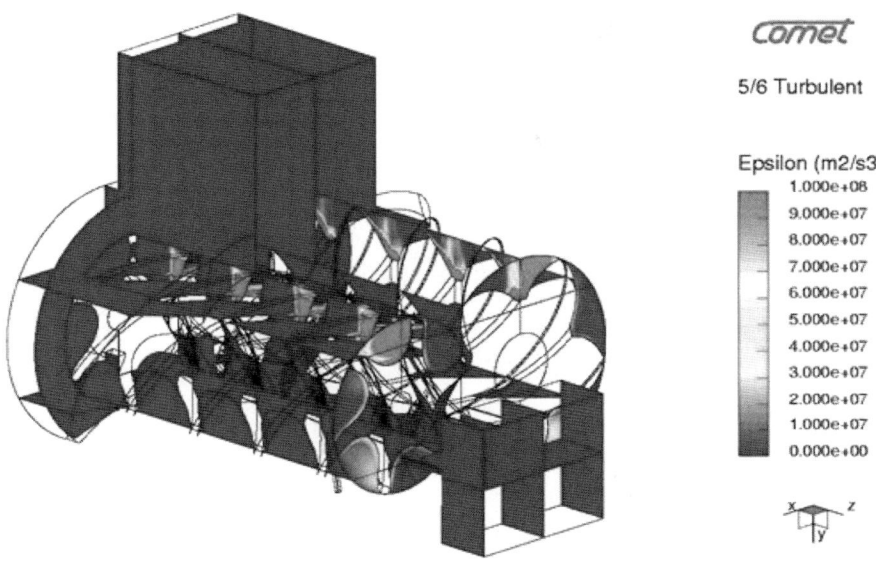

Figure 4-30 Dissipation rate within the screw compressor

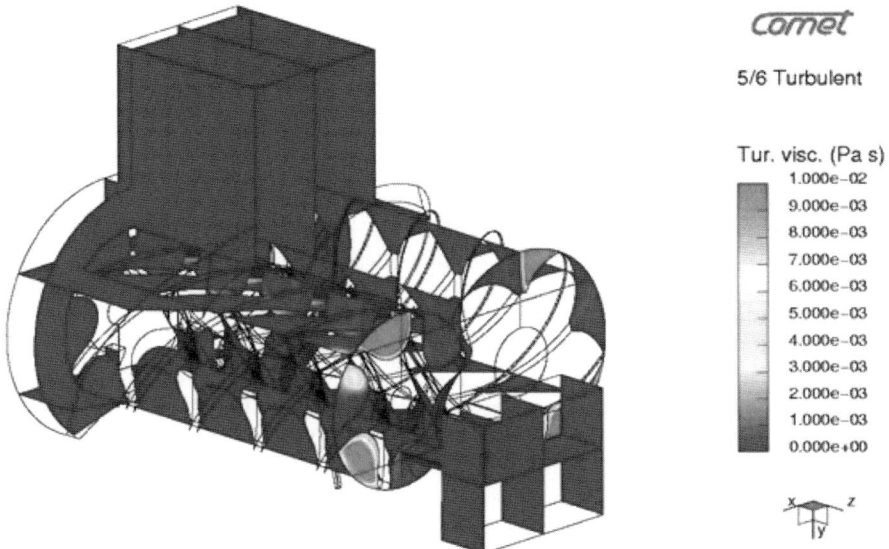

Figure 4-31 Turbulent viscosity within the screw compressor

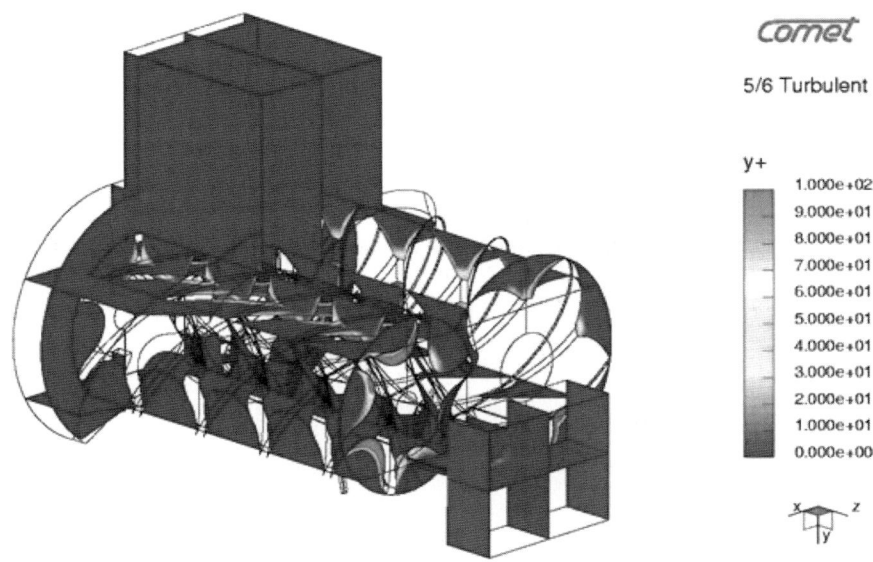

Figure 4-32 Dimensionless distances from the wall within the compressor

The results in all these diagrams are presented in horizontal sections through the blow hole areas on the suction and discharge side of the compressor, in vertical sections through the rotor axes and in cross sections at suction and discharge. The kinetic energy of turbulence, dissipation, turbulent viscosity and y+ are all high for the lobes exposed to the suction domains. All these gradually die out towards discharge. The dissipation rate is extremely high in the clearance gaps between the rotors, as shown in Figure 4-30, while in the other domains it is significantly lower. On the other hand, y+ is small in the clearance gaps while in the main domains at suction it has higher values, as shown in Figure 4-32.

4.3.5 The Influence of the Mesh Size on Calculation Accuracy

Most calculations in this book are presented for numerical meshes with an average number of 30 cells along one interlobe and a similar number of time steps selected for the rotor to rotate between two interlobe positions. The numerical mesh for the compressor in this example consists of about 450,000 cells of which About 322,000 numerical cells define the rotor domains. This was a convenient number of cells to use with a PC computer with an ATHLON 800 processor and 1GB of RAM, which was used for this study. Although the results obtained on that mesh appeared to be satisfactory and agreed well with the experimental data, an investigation of the influence of the mesh size on the calculation accuracy had to be conducted. For that reason, two additional meshes were generated for the same compressor. A smaller one was generated with 20 points along the rotor interlobe, which gave 190,000 cells on both rotors while the other compressor parts were mapped with almost the same number of cells as originally. The overall number of numerical cells was about 353,000. A lower number of cells on the rotors results in a geometry, which does not follow the rotor shape precisely, and the interconnection between rotors would possibly become inappropriate. This number of numerical cells is probably the lowest for which reliable results can be obtained. The largest numerical mesh generated for this investigation consists of 45 numerical cells along the rotor interlobe. That gave 515,520 cell on the rotors and 637,000 cells for the entire compressor domain. This was the biggest numerical mesh that could be loaded into the available computer memory without disc swapping during the solution. These three numerical meshes are presented in Figure 4-33 in the cross section perpendicular to the rotor axes.

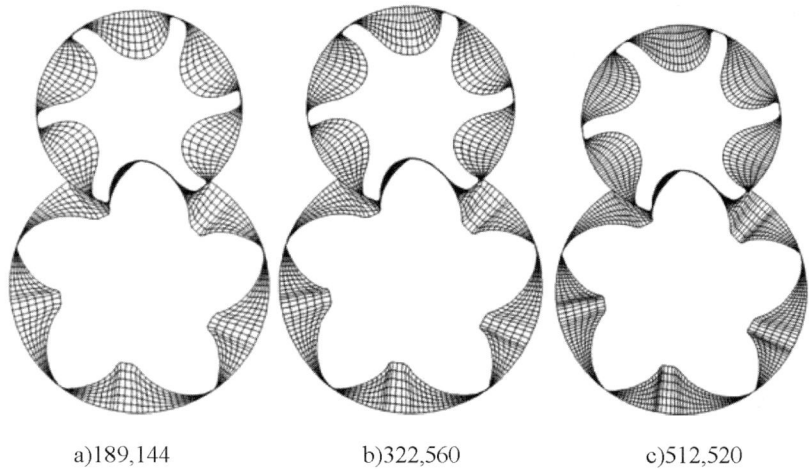

a) 189,144   b) 322,560   c) 512,520

Figure 4-33 Three different mesh sizes for the same compressor

The results of the calculations are presented in Figure 4-34 in the form of a pressure-angle diagram, and in Figure 4-36 as a discharge flow-angle diagram. The first diagram shows how the calculated working pressures for all three investigated mesh sizes agree with the measurements. The lowest number of cells gives the highest pressure in the working chamber and vice versa. As a result of that, the consumed power is changed slightly, from 42 kW obtained with the smallest mesh to slightly less then 41 kW for the largest mesh. The difference between the two is less then 3%. This situation is shown in Figure 4-35. The diagram shows the largest difference within the cycle to be in the discharge area of the compressor. Some difference is also visible in the middle area of the diagram which seems to be a consequence of the leakage flows obtained with smaller meshes between the rotors. In that area, the mesh is probably too coarse to capture all the oscillations which appear in the flow.

116 4 Applications

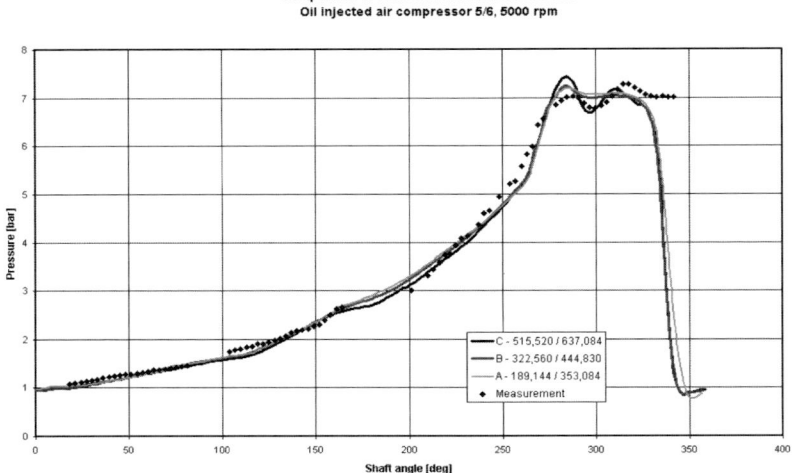

Figure 4-34 P-alpha diagrams for three different mesh sizes

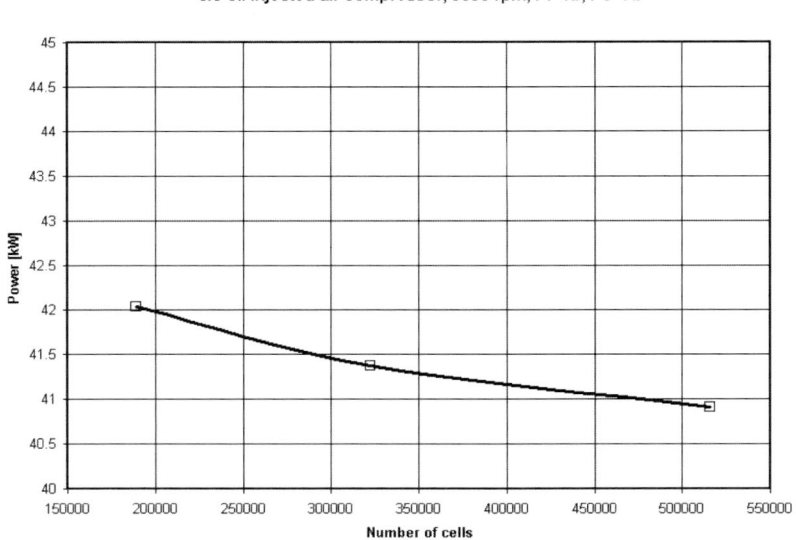

Figure 4-35 Compressor power calculated with three different mesh sizes

4.3 Flow in an Oil Injected Screw Compressor   117

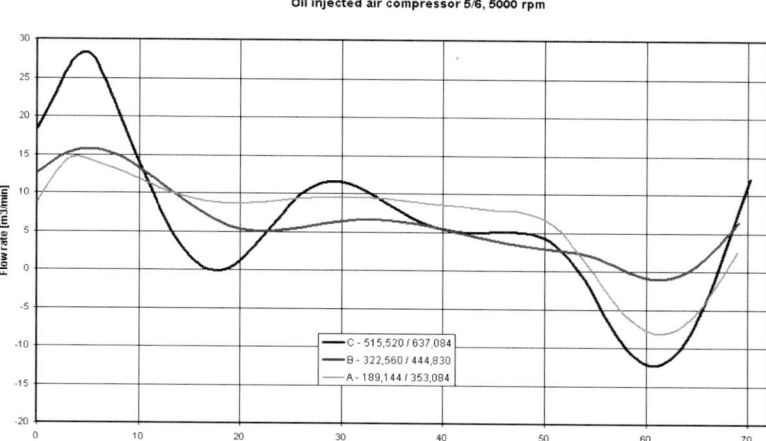

Figure 4-36 Discharge flow rates for different mesh sizes

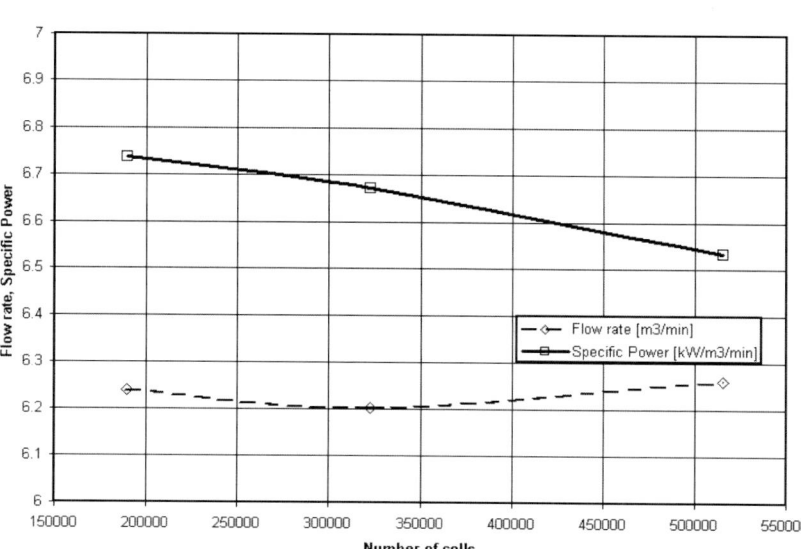

Figure 4-37 Integral flow rate and Specific power obtained with different mesh sizes

Diagrams of discharge flow as a function of rotation angle are given in Figure 4-36. The coarser mesh shows less oscillation in the flow then the finer meshes. However, the mean value of the flow remained the same for all three mesh sizes, as shown in Figure 4-37. Specific power is calculated from the values obtained previously. It shows a slight fall in value as the number of computational cells is increased.

The results obtained with the three different mesh sizes for the compressor investigated here give the impression that the calculation conducted for the compressor on an average size of the mesh with 25 to 30 numerical cells along the rotor interlobe is sufficiently accurate.

4.4 A Refrigeration Compressor

This example is given for an oil injected ammonia refrigeration compressor with 5 lobes on the male rotor and 6 lobes on the female one. The distance between the rotor axes is 108.36 mm, the diameter of the male rotor is 144.43 mm and the female rotor diameter is 121.92 mm. The rotor length is 200 mm, which gives a length over diameter ratio $l/d = 1.384$. The clearances are distributed on the compressor rotors such that the mean value is 100 µm. The compressor rotates at 5000 rpm and works between 2 bar at suction and 7 and 9 bar at discharge. A cross section through the compressor rotors is given in Figure 4-38.

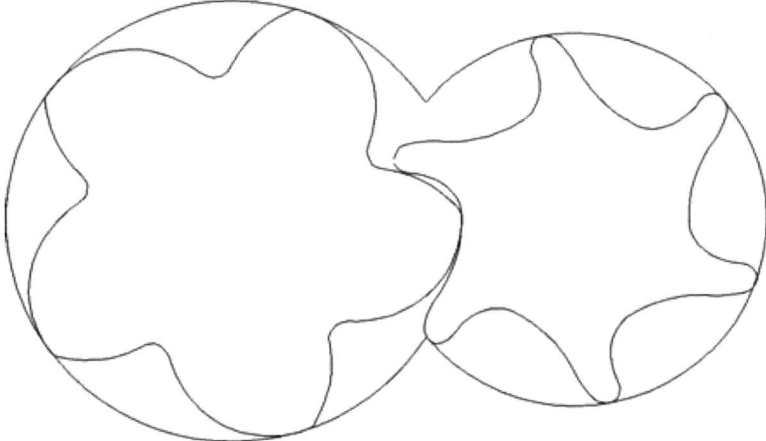

Figure 4-38 Cross section of the ammonia refrigeration compressor rotors

4.4.1 Grid Generation for a Refrigeration Compressor

The numerical mesh of this compressor consists of 256,690 numerical cells generated with 20 computational points distributed along one interlobe. Such a low number of cells has been selected because the geometry of the compressor appeared to be convenient for grid generation. That is because a large radius was put on the female rotor tip, which gives less curvature of the rotor lobes. Also, the rotor lobes are relatively shallow which is a consequence of the profile generation method used for these rotors. It is explained in more detail by *Rinder* (1984). All these factors were convenient for grid generation with only small values needed for the geometrical criteria used in boundary adaptation on the rotors. The mesh was checked both visually and numerically and the appropriate connections between the mesh blocks were established. Only 20 time steps for one interlobe rotation were needed for reliable results.

4.4.2 Mathematical Model of a Refrigeration Compressor

Ammonia is a natural working fluid with favourable features. In this example, the fluid properties of ammonia are calculated from an equation of state of a real fluid implemented in the numerical solver through user functions, as is explained in section 2.2. An equation for liquid concentration was included in the calculation, in addition to the standard momentum, energy, mass, space and oil concentration equations. By this means, fluid phase change was accounted for in the model. However, oil is regarded as an inert fluid which did not dissolve in the main fluid. Such an assumption appeared to be appropriate in this case because the production of liquid within the compression process was negligible. This was confirmed by the values of liquid refrigerant concentration which were practically zero throughout the complete domain.

4.4.3 Three Dimensional Calculations for a Refrigeration Compressor

Results of the three dimensional calculation are presented in the form of the velocity field, pressure distribution and oil concentration in two different cross sections, one normal to the rotor axis and in another horizontal section through the blow hole area. Based on these, integral parameters are calculated for this compressor in the form of pressure, torque and delivered flow.

Figure 4-39 shows an axial section through the rotors of the oil injected ammonia compressor. The unshaded area on the left of the figure outlines the male rotor while that on the right shows the female rotor. The section is taken through the high pressure blow hole area with the suction port at the top of the figure and the discharge opening at the bottom. The rotor profile allows a large blow hole area through which a substantial amount of fluid leaks back to the domain at lower pressure. The pressure distribution is shown in this figure together with the velocity vectors. The oil injection port is positioned on the bottom right of the figure. The oil enters the compressor through that port at relatively high speed. High velocities indicate a substantial leakage in the blow hole. The discharge pressure in this case was 9 bar. In the figure, the values of the relative pressure are given.

The oil concentration is presented in Figure 4-40, together with the velocity field. The dark shades on the bottom right of the figure indicate a high oil concen-

120 4 Applications

tration at the injection point. Oil is distributed across the entire compressor with the higher concentration towards the discharge port. However, some oil, leaks back, together with the gas, to the lower pressure domains through both the clearances and the blow hole. This situation is attained when the oil injection port is positioned properly and is shown in Figure 4-41.

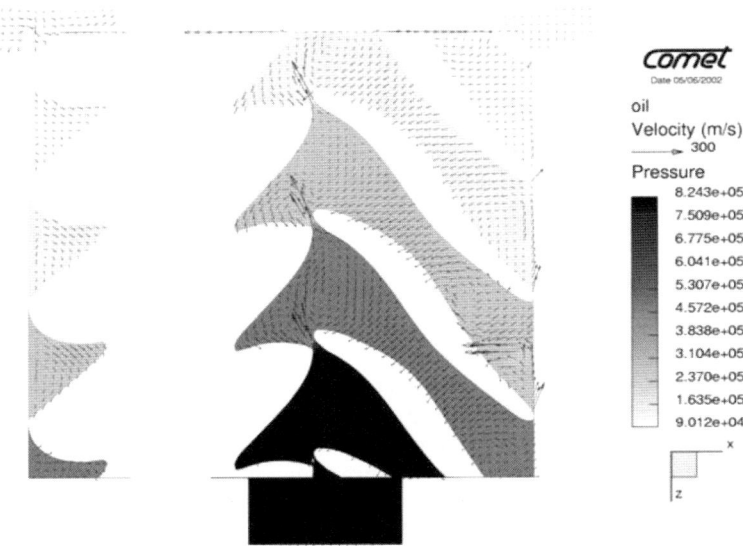

Figure 4-39 Pressure distribution and velocity field in the ammonia screw compressor

Figure 4-40 Oil concentration and velocity field for ammonia screw compressor

4.4 A Refrigeration Compressor

The position of the oil port is important for the compressor efficiency. It is usually located at the place in the compressor where temperature of the gas has the same value as that of the injected oil and where the pressure in the compressor chamber is lower than the pressure in the oil tank. If these criteria are met, the oil concentration would be as in Figure 4-41. A substantial amount of oil is then distributed along the compressor for lubrication and cooling.

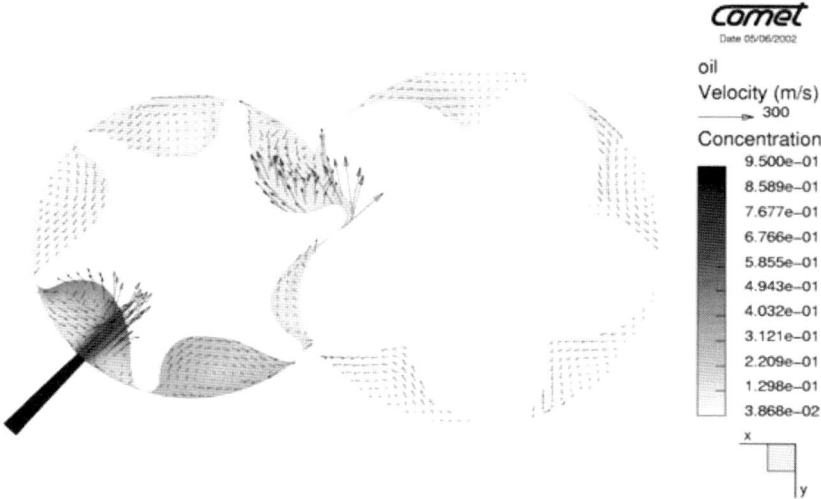

Figure 4-41 Oil concentration in the compressor with proper position of the oil injection port

Figure 4-42 Oil concentration in the compressor with improper position of oil injection port

However, if the oil port was rotated by only 45°, as shown in Figure 4-42, the oil would not flood the compressor. This could lead to an excessive rise in temperature and decrease of efficiency. Also, a low amount of oil would not be sufficient for lubrication and seizure of the rotors is then possible. Although the velocity distribution is different for the two cases, it seems that the velocity in the clearance gaps has not changed much.

A pressure versus angle diagram is presented in Figure 4-43 for discharge pressures of 7 and 9 bar. The pressure gradient in the last phase of the compression process for 7 bar discharge is lower than is the case for the higher discharge pressure. This is the consequence of leakage through the large blow hole area.

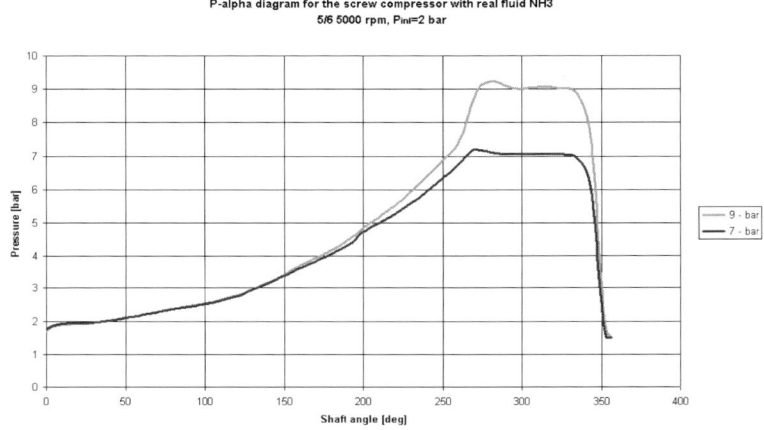

Figure 4-43 Pressure angle diagram of an oil injected ammonia screw compressor

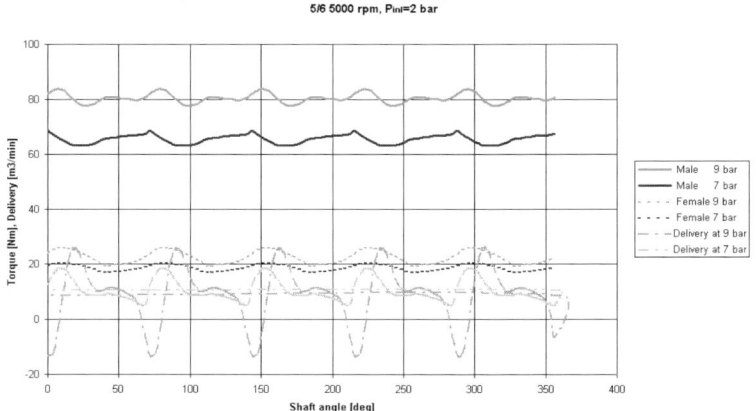

Figure 4-44 Torque and discharge flow diagrams of an ammonia screw compressor

Figure 4-44 presents the torque distribution between the male and female rotors and the discharge flow for both 7 and 9 bar discharge pressures. The torque transmitted to the female rotor is large in this compressor and is more then 20% of the male rotor torque. The light female rotor is therefore more likely to deform and possibly seize. The discharge flow rate is presented in the same diagram. Larger oscillations in the compressor discharge flow are recorded during the compressor cycle with a higher discharge pressure. The fluid flow is lower at the higher pressure, which is confirmed by the mean values of the discharge flow presented in the figure. The reason for lower flow at higher pressures is due to the higher leakage flow which is consequence of larger pressure difference through both the blow hole areas and the clearances.

4.5 Fluid-Solid Interaction

Efforts are continually being made to produce screw compressors with smaller clearances in order to reduce internal leakage. However, since the compression process induces large pressure and temperature differences across the rotors, they deform. A reliable method of estimating the interaction between fluid flow parameters and rotor deflection is thus needed in order to minimise clearances while avoiding contact between the rotors and the casing. A 3-D mathematical procedure is presented here to generate a numerical grid comprising both solid and fluid domains. This can be used to calculate the fluid flow and compressor structural deformation simultaneously by means of a suitable commercial numerical solver. Simulation results demonstrate the effect on compressor performance of changes in working clearances, caused by rotor deformation.

4.5.1 Grid Generation for Fluid-Solid Interaction

Grid generation of screw compressor geometry is a necessary preliminary to CCM calculation. Firstly, it defines spatial domains that represent the metal material inside the rotors and the fluid passages outside the rotors. These are determined by the rotor profile coordinates and their derivatives and are obtained by means of the rack generation procedure.

By use of the methods already described, an algebraic grid generation method for screw compressor fluid flow is applied, based on a transfinite interpolation procedure. This includes stretching functions to ensure grid orthogonality and smoothness. In that case, the compressor spatial domain is divided into a number of sub-domains, which allow the generation of a fully structured numerical grid of discrete volumes. A composite grid is then made of several blocks patched together and based on a single boundary fitted co-ordinate system. Block structured grids allow easier mapping for complex geometries. Two basic block topologies are used for screw compressor grid generation, namely polyhedral blocks and O-meshes as shown in Figure 3-5.

A feature associated with the interaction of the fluid flow and solid structure is that the numerical grid for both, the fluid around the rotors and the rotors themselves is generated simultaneously in a single, fully structured block. This allows distortions to change the interlobe, radial and end clearances and accounts for them in the fluid flow calculation. The number of points required to define the rotor geometry accurately in this case is generally not very large. However, if a sliding interface between the solid and fluid is applied, the number of points needed may be so large that a numerical mesh, thus formed, cannot be used. One means of resolving this problem, which combines accuracy with fast solution, is to keep the number of computational cells as low as possible and to modify the distribution of points according to the local requirements. An additional reason for this approach is that the principal dimension of a screw compressor chamber may vary from as little as 30 micrometers to tens of millimetres. It is therefore not unusual for a grid length scale ratio to exceed 500. Since the number of cells in the radial direction is kept constant throughout the compressor solid section, as well as in the flow chamber and in the gaps, the ratio between the circumferential and radial dimensions of the cell can easily become unacceptable. However, the same number of cells can form a suitable grid if the boundary is adapted carefully to keep the grid aspect ratio as uniform as possible.

An equi-distribution procedure is therefore applied to the boundary regions between the rotor and its fluid domain. It requires the product of the grid spacing and a 'weight function' to be constant. The weight function is based on an adaptation variable, which can be selected according to geometry requirements. The adaptation procedure is carried out in four major steps; namely: a) selection of adaptation variables, b) evaluation of integral adaptation functions, c) calculation of new transformed coordinate and d) obtaining new physical coordinate by inverse interpolation.

Once a satisfactory point distribution on the rotor boundary is achieved, the distribution of points on the opposite sliding interface boundary, must be obtained for convenient generation of the numerical points in the interior of the domain. That boundary is mapped with the same number of points as the previous one so the cells that are formed from the appropriate nodes on both boundaries are always regular. For that reason, a special procedure was developed to transform the physical domain into a simplified computational domain in which the adaptation is performed.

4.5.2 Numerical Solution of the Fluid-Solid Interaction

The compressor flow and the structure of compressor parts are fully described by the mass averaged conservation equations of fluid continuity, momentum equations for fluid and solid body, energy and space, which are accompanied by the turbulence model equations and an equation of state. In the case of multiphase flow, a concentration equation is added to the system. The numerical solution of such a system of partial differential equations is then made possible by inclusion of constitutive relations in the form of Stoke's, Fick's and Fourier's laws for the fluid momentum, concentration and energy equations respectively and Hooke's law for the momentum equations of a thermo-elastic solid body.

All these equations are conveniently written in the form of a generic transport equation (2.61). The meaning of source terms, for the fluid and solid transport equations is given in *Table 2-1*.

The resulting system of partial differential equations is discretised by means of a finite volume method in a general Cartesian coordinate frame. This method enhances the conservation of the governing equations while at the same time enables a coupled system of equations to be solved simultaneously for both the solid and fluid regions.

This mathematical scheme is accompanied by boundary conditions for both the solid and fluid parts. The compressor was positioned between the two relatively small suction and discharge receivers. By this means, the compressor system is separated from the surroundings by adiabatic walls only. It communicates with its surroundings through the mass and energy sources or sinks placed in these receivers to maintain constant suction and discharge pressures.

The effects of oil injected in the working chamber are calculated based on the Euler-Lagrange method. In that case, the air is regarded as a background fluid for which all transport equations are solved, while the oil is considered as a dispersed phase for which only a concentration equation is solved. Interaction between the phases in the form of drag forces and heat transfer is calculated based on an experimentally derived mean droplet diameter. The solid part of the system is constrained by both Dirichlet and Neuman boundary conditions through zero displacement in the restraints and zero tractions elsewhere. The temperature and displacement from the solid body surface are boundary conditions for the fluid flow and vice versa. The connection between the solid and fluid parts is therefore determined explicitly.

The numerical grid was applied to a commercial CCM solver to obtain the distribution of the pressure, temperature, velocity and density fields throughout the fluid domain as well as deformation and stress in the solid compressor elements. Integral parameters of screw compressor performance were calculated, based on the solution of these equations,.

4.5.3 Presentation and Discussion of the Results of Fluid-Solid Interaction

The interaction between fluid flow and rotor deformation is analysed for a screw compressor shown in Figure 4-14. The Rotor profiles are of the 'N' type with a 5/6 lobe configuration. The rotor outer diameters are 127 and 101 mm, for the male and female rotors respectively, and their centre distance is 90 mm. The rotor length to diameter ratio is 1.65. A numerical mesh for the test case in this study comprises 513,617 numerical cells of which 162,283 cells represent the solid part of the rotors, 189,144 other cells are mapped on the fluid parts between the rotors while the rest are numerical cells of the suction and discharge domains, which include both the suction and discharge ports and the oil openings.

126 4 Applications

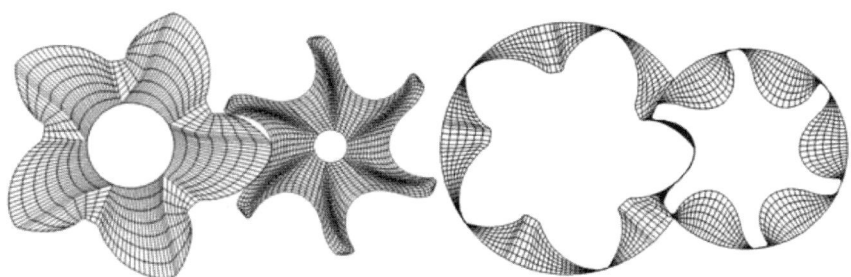

Figure 4-45 Numerical mesh for rotors (left) and their fluid parts (right)

A cross section through the numerical mesh for rotors and their fluid paths is presented in
Figure 4-45.

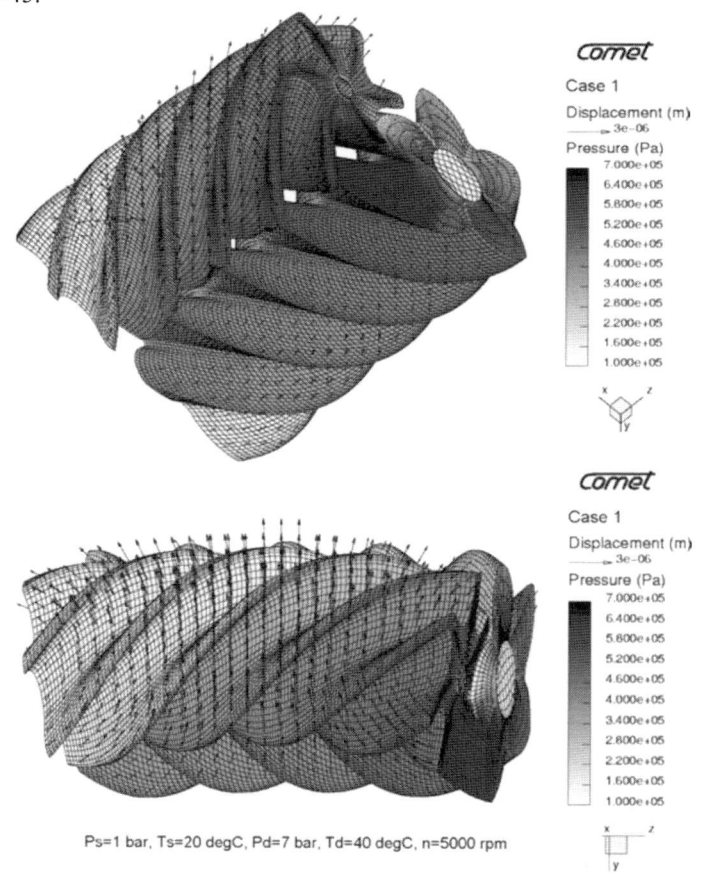

Ps=1 bar, Ts=20 degC, Pd=7 bar, Td=40 degC, n=5000 rpm

Figure 4-46 Displacement vectors and the acting pressure on deformed rotors for oil injected compressor

4.5 Fluid-Solid Interaction

Grid and other control parameters generated by the interface were applied to Comet, a commercial CMM solver of CD-Adapco. The solver is based on the finite volume method and adapted for the simultaneous multi domain solution of the governing laws. Depending on the selected set of equations, alternative differencing schemes and relaxation factors were used. The results obtained are presented for three different applications, namely, for an oil-injected air compressor of moderate pressure ratio, a dry air compressor, in which the pressure ratio is low, due to discharge temperature restrictions, and a high pressure oil flooded compressor.

The calculations were carried out on a computer powered by an Athlon 800 MHz processor and 1 GB memory. Compressor rotation was simulated by means of 24 time steps for one interlobe rotation. This was equivalent to 120 time steps for one full rotation of the male rotor. The time step length was synchronised with a compressor speed of 5000 rpm. An error reduction of 4 orders of magnitude was required, and achieved in approximately 50 outer iterations at each time step, each of which took approximately 30 minutes of computer time. The overall compressor parameters such as torque, volume flow, forces, efficiencies and compressor specific power were then calculated.

In case 1, the oil flooded air compressor works between 1 bar and 20 °C suction and 7 bar discharge. A substantial amount of oil was injected in the compressor in order to keep the discharge temperature as low as possible. The average temperature at the compressor exit of 40°C was retained throughout the compressor working cycle. Consequently, rotor deformation was caused mainly by the pressure field in the compressor working chamber.

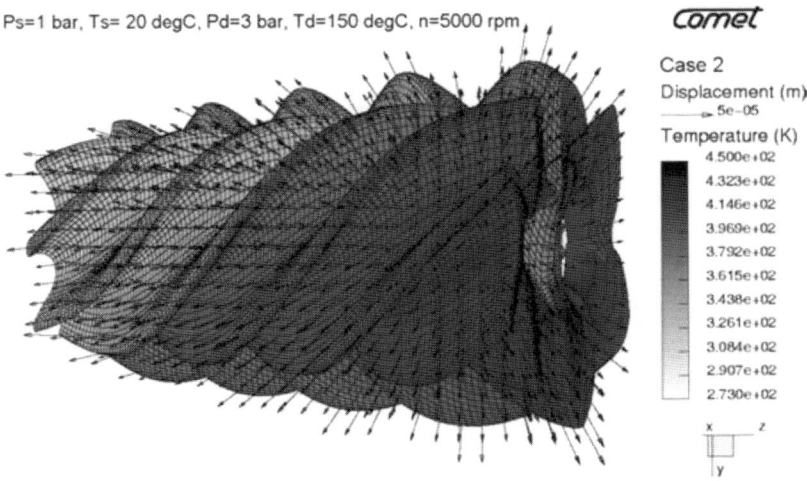

Figure 4-47 Rotor displacement vectors and temperature distribution for an oil free compressor

Two views of the rotors for that condition are presented in Figure 4-46, one from the bottom of the rotors and the other one from the female rotor side. Pressure forces push the rotors apart and bend them in space, which is visible from the top diagram. By this means, the leakage path through the sealing line between the rotors is biggest on the discharge side where the pressure ratio between the two rotor sides is the greatest. Also, rotor bending increases the clearance gap between the rotor and the housing on the discharge side of the compressor. The female rotor is weaker and therefore is deformed more than the male rotor. The highest recorded deformation in this case was in the range of only a few micrometers. In order to visualise the rotors in the deformed state, the distortion has been magnified 20,000 times while other physical and geometrical values are kept at their original scale.

Case 2 is an oil free air compressor of the same rotor and housing arrangement as in Case 1. The compressor works between suction conditions of 1 bar and 20 °C and a discharge pressure of 3 bar. Due to the lack of oil cooling, the temperature rise in the compressor is much greater, so the temperature at the discharge port has an average value of 150°C. Such working conditions cause a completely different rotor deformation to that in Case 1. This is presented in Figure 4-47. The deformation caused by the pressure difference is negligible compared to that caused by the temperature change. The fluid temperature in the immediate vicinity of the solid boundary changes rapidly as shown in the bottom diagram. However, the temperature of the rotor pair is lower due to the continuous averaging oscillations of pressure and temperature in the neighbouring fluid. This is shown in the top diagram of Figure 4-47, where the temperature distribution is given in cross section for both the fluid flow and the rotor body. The deformation shown in the bottom of the figure demonstrates the enlargement of the rotors in the discharge area. This is more than an order of magnitude higher than in the case of the oil flooded compressor. It reaches $50 \mu m$ in the discharge section of the rotors and thus reduces the working clearances between the rotors by the same amount. By this means, the leakage flow in the most critical areas is reduced. However, in a design of the dry screw compressor, sufficient initial clearance should be allowed to prevent rotor seizure caused by temperature distortion. The deformations in Figure 4-47 are magnified 1,500 times in order to make them visible.

4.5 Fluid-Solid Interaction   129

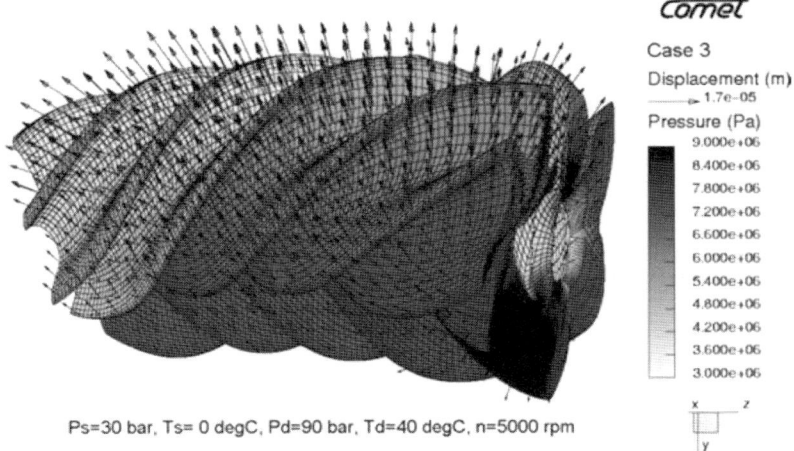

Figure 4-48 Deformations of a high pressure oil injected compressor

Case 3 represents a high pressure oil injected compressor. An example is given here for a CO_2 compressor with suction conditions of 30 bar and 0°C and a discharge pressure of 90 bar. The discharge temperature was 40°C. In this case, the large pressure difference caused higher rotor deflections than in Case 1, as shown in Figure 4-48. The highest deformation was in excess of 15μm, which is of the same order of magnitude as was found in the case where temperature deformation was dominant. The deformation pattern of the rotors is similar to that in Case 1 but only slightly enlarged at the discharge side.

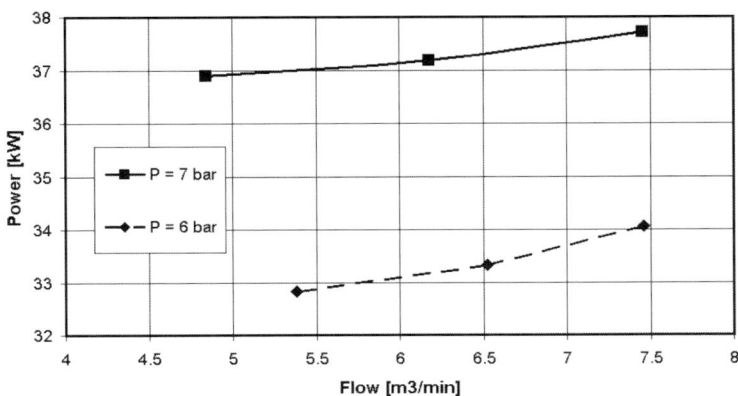

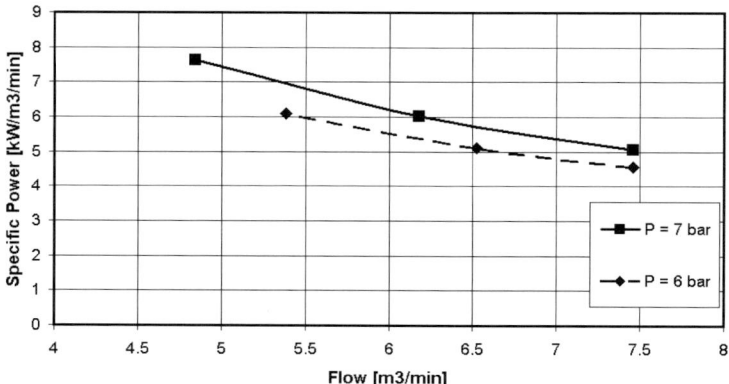

Figure 4-49 Influence of the rotor deflection to the integral compressor parameters

The influence of the rotor deformation on the integral screw compressor parameters caused by the change in clearance is given in Figure 4-49. The reduction of rotor clearances due to the enlargement of the rotors caused by temperature dilatations results in an increase in both, the compressor flow and power input. However the flow increase is relatively larger than that of the power and hence results in a decrease in specific power, or more conventionally, an increase in efficiency, as shown in the diagram. However, the rotor deflections, caused by the pressure, enlarge the clearances. For a moderate compressor pressure, the clearance gap is enlarged only slightly and hence has only a negligible influence on the delivery and power consumption. In the case of high working pressures as, for example, in CO_2 a refrigeration application, the rotors deform more and the decrease in the delivery and rise in specific power becomes more pronounced.

5
Conclusions

The analytical methods described in this work should be regarded as indicators of the scope for improvement that can yet be made in twin screw compressor and expander design rather than final solutions. Thus, the procedures used are still in the process of validation by continuing experiment. However, while validation proceeds, these same methods are being extended further to attempt to use the solid-fluid interaction effects to model noise generation in screw compressors and use this to eliminate unwanted effects at the design stage. In addition, a management system is being developed to integrate one-dimensional flow models and three-dimensional flow models with a CAD system in an interactive manner. By this means a compressor can be designed, initially using a simple one-dimensional flow approach. This can then be improved by applying three-dimensional modelling where it is most needed and any changes, thus generated, will be automatically incorporated in the original analysis to show how such design changes will affect the final performance. In the long run, similar as in internal combustion engine modelling, it may be possible to produce a virtual compressor, with full performance simulations prior to prototype manufacture.

A
Models of Turbulent Flow

The Navier-Stokes equations can be used to describe any fluid flow, including turbulent flows. However, their direct numerical solution for turbulent flows requires a mesh with spacing smaller than the length scale of the smallest turbulent eddies at which the energy is transformed to heat and time steps smaller than the smallest time scale of the turbulent fluctuations. The average length scale of the smallest eddies in positive displacement machines is expected to be of the order of $10\mu m$ while their time scale is of the order of milliseconds, *Lumley* (1999). Useful solutions on this scale are beyond the scope of existing computer technology.

The alternatives are either large eddy simulation, in which only the largest unsteady motions are resolved and the rest are modelled, or the use of Reynolds averaged Navier-Stokes equations (RANS), obtained by the use of a statistical description of turbulent motion, formulated in terms of averaged quantities. The most popular is the averaging for flows with constant density, whereby each dependent variable is expressed as the sum of its mean, or time-averaged value $\overline{\phi}$, and a fluctuating component ϕ'':

$$\phi = \overline{\phi} + \phi''. \tag{A.1}$$

The value $\overline{\phi}$ is averaged over a time interval, which is large enough, compared to the time scale of the turbulent fluctuations, but small with respect to the scale of other time dependant effects. For compressible flow a density-weighted, Favre averaging is used through the following definition:

$$\phi = \overline{\phi} + \phi' \tag{A.2}$$

with the time averaged value $\overline{\phi} = \overline{\rho\phi}/\overline{\rho}$, and the fluctuating component calculated as $\overline{\rho\phi'} = 0$.

Applied to the governing equations of Section 3.2.1, the averaging procedure produces a set of additional unknowns in the equations of momentum, energy and species. These unknowns are fluctuating parts in the diffusive term of the equations, while other dependent variables are considered as averaged quantities. With this approach, the stress tensor **T** in the momentum equation is substituted with its

134 A Models of Turbulent Flow

effective value $\mathbf{T}^e=\mathbf{T}+\mathbf{T}^t$, the heat flux $\mathbf{q}_h$ in the energy equation is incorporated in the effective value $\mathbf{q}_h^e=\mathbf{q}_h+\mathbf{q}_h^t$, while the effective value of diffusive mass flux in the concentration equation becomes $\mathbf{q}_{ci}^e=\mathbf{q}_{ci}+\mathbf{q}_{ci}^t$. The averaged parts are the same as in the original equations while the fluctuating parts are defined as follows:
Turbulent momentum flux, known as Reynolds stress is:

$$\mathbf{T}^t = -\overline{\rho \mathbf{v}'\mathbf{v}'} \tag{A.3}$$

Turbulent heat flux is

$$\mathbf{q}_h^t = -\overline{\rho e' \mathbf{v}'} \tag{A.4}$$

Turbulent mass flux is

$$\mathbf{q}_{ci}^t = -\overline{\rho c_i' \mathbf{v}'} \tag{A.5}$$

These new unknown values are accompanied by a turbulence model which provides correlations of the fluctuations in terms of the mean quantities, in order to be incorporated into the governing equations of Section 3.2.1. The popular turbulence models are eddy-viscosity models based on the analogy between turbulent and viscous diffusion. By the use of this model, equations (A.3) to (A.5) become:

$$\begin{aligned}\mathbf{T}^t &= 2\mu_t \dot{\mathbf{D}} - \frac{2}{3}(\mu_t \operatorname{div} \mathbf{v} + \rho k)\mathbf{I}, \\ \mathbf{q}_h^t &= \kappa_t \operatorname{grad} T, \\ \mathbf{q}_{ci}^t &= \rho D_{i,t} \operatorname{grad} c_{ui}.\end{aligned} \tag{A.6}$$

The effect of turbulence is introduced through the turbulent diffusivity, viscosity and conductivity as given in (A.8). These are fluctuating parts of the effective diffusivity, viscosity and conductivity values:

$$\begin{aligned}D_{i,\mathit{eff}} &= D_i + D_{i,t}, \\ \mu_{\mathit{eff}} &= \mu + \mu_t, \\ \kappa_{\mathit{eff}} &= \kappa + \kappa_t,\end{aligned} \tag{A.7}$$

defined respectively in the next set of equations:

$$\rho D_{i,t} = \frac{\mu_t}{\sigma_{ci}}, \qquad \mu_t = \rho C_\mu \sqrt{k} L, \qquad \kappa_t = \frac{\mu_t C_p}{\sigma_T}. \qquad (A.8)$$

Here, $\sigma_{ci}, C_\mu, \sigma_T$ are empirical coefficients reasonably similar for all RANS turbulence models. They are usually C_μ =0.09, $\sigma_{ci} = \sigma_T$ =0.9. In the previous equation k is the kinetic energy of turbulence and L is a length scale. Turbulent kinetic energy is defined as:

$$k = \frac{1}{2} \overline{\mathbf{v}' \cdot \mathbf{v}'} \qquad (A.9)$$

Turbulent diffusivity and conductivity in (A.8) are directly estimated from the turbulent viscosity.

Two RANS models are used in compressor calculation, the Zero-Equation model and the Standard k-ε two-equation model. More details on turbulence phenomena can be found in such works as *Ferziger and Peric* (1995) and *Wilcox* (1993).

Zero Equation Model

Turbulence may be characterized by its kinetic energy k or by the turbulent velocity $q = \sqrt{2k}$, and the length scale L. From equation (A.8) the turbulent viscosity for the zero-equation model is:

$$\mu_t = C_\mu \rho q L \qquad (A.10)$$

The kinetic energy of turbulence in this model is determined from the mean velocity field using:

$$q = L \frac{\partial v}{\partial y} \qquad (A.11)$$

where L is a given function dependent on the coordinates. Accurate determination of L is difficult for separated and three-dimensional flows and, despite being simple, this model is not often used for engineering flow simulation.

k-ε Model

The difficulty in prescribing turbulence quantities suggests the use of differential equations for their estimation. Since the turbulent velocity and length scale are needed, a model is based on two such equations. The equation of the turbulent kinetic energy k usually determines the velocity scale. The equation of the length scale is derived, taking into account that, for equilibrium turbulent flows, the dissipation ε, kinetic energy k and length ratio L are related by:

$$\varepsilon \approx \frac{k^{3/2}}{L} \tag{A.12}$$

Partial differential equations for both kinetic energy and its dissipation can be derived from the Navier-Stokes equations. The final form of these equations is given as:

$$\frac{d}{dt}\int_V \rho k \, dV + \int_S \rho k (\mathbf{v} - \mathbf{v}_s) \cdot d\mathbf{s} = \int_S \mathbf{q}_k \cdot d\mathbf{s} + \int_V (P - \rho \varepsilon) \, dV \tag{A.13}$$

$$\frac{d}{dt}\int_V \rho \varepsilon \, dV + \int_S \rho \varepsilon (\mathbf{v} - \mathbf{v}_s) \cdot d\mathbf{s} = \int_S \mathbf{q}_\varepsilon \cdot d\mathbf{s} + \int_V \left(C_1 P \frac{\varepsilon}{k} - C_2 \rho \frac{\varepsilon^2}{k} - C_3 \rho \varepsilon \, \text{div} \mathbf{v} \right) dV$$

where k is the turbulent kinetic energy, ε is its dissipation, while P is production of turbulence energy given by:

$$P = \mathbf{T}^t : \text{grad } \mathbf{v} = 2\mu_t \dot{\mathbf{D}} : \dot{\mathbf{D}} - \frac{2}{3}(\mu_t \text{div } \mathbf{v} + \rho k) \text{div } \mathbf{v} \tag{A.14}$$

Diffusion fluxes in (A.13) are:

$$\mathbf{q}_k = \left(\mu + \frac{\mu_t}{\sigma_k}\right) \text{grad } k, \qquad \mathbf{q}_\varepsilon = \left(\mu + \frac{\mu_t}{\sigma_\varepsilon}\right) \text{grad } \varepsilon \tag{A.15}$$

while, after implementing (A.12), the turbulent viscosity is:

$$\mu_t = C_\mu \rho \frac{k^2}{\varepsilon} \tag{A.16}$$

The constants are derived experimentally and they are: $C_\mu = 0.09$, $\sigma_k = 1$, $\sigma_e = 1.3$, $C_1 = 1.44$, $C_2 = 1.92$, $C_3 = -0.33$.

Implementation of the k-ε turbulence model is relatively simple. All the model equations have the same form as the main flow equations if the viscosity is replaced by the effective viscosity (A.7). Equations for the kinetic energy of turbulence and its dissipation are solved separately, after the main equations are solved, by use of the k and ε values from the previous iteration. The main reason for that is the 'stiffness' of the turbulence model equations. To overcome the problem, either a finer grid could be used for the turbulence model equations than for the main equations or a local blending of numerical schemes of different order can be applied. It is also useful to under-relax the iterative method for these two quantities in order to avoid negative values of k and ε, which can lead to numerical instability.

B
Wall Boundaries

Two types of wall are applied to a screw compressor, namely, moving walls, if they bound the domain of the compressor rotors, and stationary walls in all other places. Boundary conditions on these walls are explicitly given for all equations in the numerical model. In the case of turbulent flow, dependent variables vary rapidly close to the solid boundaries and a method, which can model near-wall effects is used. If the flow is laminar, then either the values of the dependent variables or their gradients are known at the boundary. Screw compressor walls are treated as 'no-slip walls', where the fluid velocity at the wall coincides with the wall velocity. All stationary walls have zero wall velocity. The rotor wall velocities are calculated from the given rotational speed:

$$\boldsymbol{\omega}_1 = \frac{2\pi n}{60}\mathbf{k}; \qquad \mathbf{v}_{1i} = \mathbf{r}_{1i} \times \boldsymbol{\omega}_1$$
$$\boldsymbol{\omega}_2 = -\frac{z_1}{z_2}\boldsymbol{\omega}_1; \qquad \mathbf{v}_{2i} = \mathbf{r}_{2i} \times \boldsymbol{\omega}_2 \qquad (B.1)$$

Indices *1* and *2* indicate the male and female rotors respectively. z_1 and z_2 are the number of teeth on the rotors. $\mathbf{r}_{1i}$ and $\mathbf{r}_{2i}$ are the vectors of the boundary points on the male and female rotors in the absolute coordinate system while ω_1 and ω_2 are the angular velocities of the male and female rotors.

If the numerical grid is coarse, large velocity variations in the near-wall region have to be interpolated in order to obtain realistic values of the shear stress. The mode of interpolation is dependent on the wall functions. These are based on a logarithmic velocity profile. Thus, the viscosity near the wall is replaced by the value μ_w, which is determined from the logarithmic velocity profile as:

$$\mu_w = \frac{y^+}{u^+}\mu; \qquad u^+ = \begin{cases} y^+ & \text{for} \quad y^+ < y_v^+ \\ \frac{1}{K}\ln(\mathcal{E}y^+) & \text{for} \quad y^+ \geq y_v^+ \end{cases} \qquad (B.2)$$

where ε is the wall roughness, which is constant for a logarithmic profile. K=0.41 is the von Karman constant, y^+ is a nondimensional distance from the wall and y_v^+ is the viscous sub-layer thickness.

If the zero-equation turbulence model is applied, then

$$y^+ = \frac{1}{K}\frac{\mu+\mu_t}{\mu}. \tag{B.3}$$

For the k-ε model of turbulence this distance is:

$$y^+ = \frac{\rho\sqrt[4]{C_\mu}\sqrt{k_P}\delta_{nP}}{\mu}. \tag{B.4}$$

In the previous equation, k_P is the kinetic energy of turbulence at the centre of the boundary cell, while σ_{nP} denotes the normal distance from the centre of the boundary cell to the wall.

The viscous sub-layer thickness is defined as the larger root of the equation:

$$y_v^+ = \frac{1}{K}\ln(\varepsilon y_v^+) \tag{B.5}$$

The boundary condition for the continuity equation is the mass flux through the wall, which is equal to zero. This condition implies zero gradient for the pressure correction in the direction normal to the wall. Thus pressure is always extrapolated from inside the solution domain.

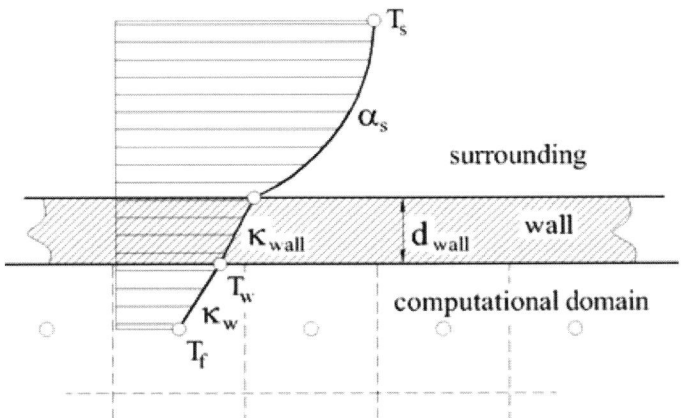

Figure A-1 Wall thermal resistance

At the wall boundaries mixed boundary conditions for the energy equation may be applied. There are no walls in the screw compressor that are adiabatic or through which the flux rate is known. The temperature on the wall is generally not known but the temperature of its surroundings, T_s, is given. If the heat transfer coefficient between the surroundings and the wall can be estimated, the wall temperature T_W

can easily be calculated from its surrounding temperature and the wall thermal resistance coefficient k_w, which is defined as:

$$k_w = \frac{d_{wall}}{\kappa_{wall}} + \frac{1}{\alpha_s} \tag{B.6}$$

where the wall thickness d_{wall}, the thermal conductivity κ_{wall} and the heat transfer coefficient between the wall and the surrounding α_s are defined in Figure A-1.

For turbulent flow, when the mesh is too coarse to resolve large temperature variations in the near wall region, interpolation based on the logarithmic temperature profile is applied. The heat flux can be obtained by use of the modified thermal conductivity in the near wall region on Figure A-1, which has the following form:

$$\kappa_w = \begin{cases} \kappa & \text{for} \quad y^+ < y_T^+ \\ \dfrac{y^+}{u^+ + \boldsymbol{P}_T} \dfrac{\mu C_p}{\sigma_T} & \text{for} \quad y^+ \geq y_T^+ \end{cases} \tag{B.7}$$

where $\boldsymbol{P}_T$ is the viscous sub-layer resistance factor defined by the Prandtl number Pr and the turbulent Prandtl number σ_T as:

$$\boldsymbol{P}_T = 9.24 \left[\left(\frac{\text{Pr}}{\sigma_T} \right)^{0.75} - 1 \right] \left[1 + 0.28 \exp\left(-0.007 \frac{\text{Pr}}{\sigma_T} \right) \right] \tag{B.8}$$

The thermal sub-layer thickness y_T^+ is defined as the larger root of the non linear equation:

$$\frac{\text{Pr}}{\sigma_T} y_T^+ = \frac{1}{K} \ln\left(\varepsilon_T y_v^+ \right) \tag{B.9}$$

where ε_T is the wall roughness parameter in the logarithmic temperature profile.

The k-e turbulence model, in conjunction with the wall functions, requires that the production of turbulent kinetic energy and the dissipation rate of the turbulent kinetic energy are given in the boundary cell. These two values are calculated from the logarithmic velocity profile. According to *Ferziger and Peric* (1995), the diffusive flux of the kinetic energy through the wall is assumed to be zero.

For the governing equations of the concentration of the oil and liquid phases, Neuman boundary conditions are applied, with the diffusive flux through the wall set to zero.

C

Finite Volume Discretisation

Governing Equations

The mathematical model for screw compressor fluid flow consists of the continuity, momentum, energy, concentration, space and turbulence model equations. All of them are written in their general form as:

$$\frac{d}{dt}\int_V \rho\phi\, dV + \sum_{j=1}^{n_f}\int_{S_j}\rho\phi(\mathbf{v}-\mathbf{v}_s)\cdot d\mathbf{s} = \sum_{j=1}^{n_f}\int_{S_j}\Gamma_\phi\, grad\phi\cdot d\mathbf{s} + \left(\sum_{j=1}^{n_f}\int_{S_j}\mathbf{q}_{\phi S}\cdot d\mathbf{s} + \int_V q_{\phi V}\, dV\right) \quad (C.1)$$

The space is divided into cells of known volume and surface area. The surface and volume integrals are replaced by quadrature approximations. The spatial derivatives are replaced by an interpolation function, a time integration scheme is applied and the surface velocities, $\mathbf{v}_s$, are determined. The result is a system of algebraic equations.

The volume of a computational cell is calculated using Gauss's theorem:

$$\int_V div\,\mathbf{r}\, dV = \int_S \mathbf{r}\cdot d\mathbf{s} \Rightarrow V_{P_o} = \frac{1}{3}\sum_{j=1}^{n_f}\mathbf{r}_j\cdot \mathbf{s}_j. \quad (C.2)$$

$\mathbf{r}_j$ is the position vector of the cell face centre, $\mathbf{s}_j$ is a cell face surface vector and n_f is the number of cell faces.

Since the edges of the control volume are straight lines, the projections of the faces onto Cartesian coordinate surfaces are independent of the surface shape:

$$\mathbf{s}_j = \frac{1}{2}\sum_{i=3}^{n_j^v}\left[(\mathbf{r}_{i-1}-\mathbf{r}_1)\times(\mathbf{r}_i-\mathbf{r}_1)\right] \quad (C.3)$$

Here, n_j^v is the number of vertices in the cell-face j, and $\mathbf{r}_i$ is the position vector of the vertex i. The cell face area is independent of the choice of the common vertex $\mathbf{r}_1$.

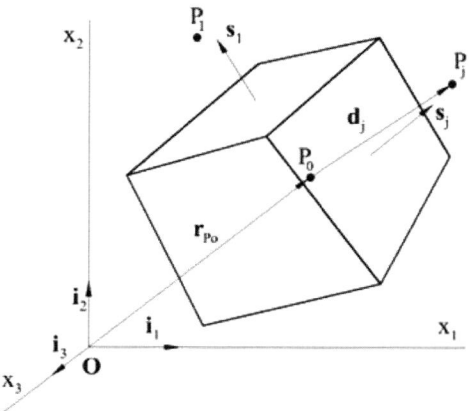

Figure A-2 Notation applied to a hexahedral control volume

The integrals and gradients are also estimated by a quadrature approximation. The simplest and the most frequently used of these is the midpoint rule approximation which is the product of the integrand at the centre of the integration domain and the surface or volume of the domain:

$$\int_{S_j} \mathbf{f} \cdot d\mathbf{s} \approx \mathbf{f}_j \cdot \mathbf{s}_j, \qquad \int_V f dV \approx f_{P_0} V_{P_0} \qquad (C.4)$$

where $\mathbf{f}$ and f are arbitrary vectors and scalars respectively.

Although the variables and fluid properties are defined at the computational nodes in the centre of the control volume, they are often needed at locations other than the cell centres. In such a case, an interpolation based on the shape function is used. A second-order linear approximation is used here.

$$\psi(\mathbf{r}) = \psi_{P_0} + (grad\psi)_{P_0} \cdot (\mathbf{r} - \mathbf{r}_{P_0}) \qquad (C.5)$$

This equation applied to the cell face centre, in the symmetric form reads:

$$\psi_j = \frac{1}{2}(\psi_{P_0} + \psi_{P_j}) + \frac{1}{2}\left[(grad\psi)_{P_0} \cdot (\mathbf{r}_j - \mathbf{r}_{P_0}) + (grad\psi)_{P_j} \cdot (\mathbf{r}_j - \mathbf{r}_{P_j})\right] \qquad (C.6)$$

The form of this equation ensures a unique approximation of the property at the centre of the cell face for control volumes on both sides of the face. $\mathbf{r}_j$ is the position vector of the cell face centre while P_0 and P_j represent the centres of the calculating and neighbouring control volume, as shown in Figure 2-2. The first term in this equation is a linear interpolation between the neighbouring cell centres, while the second term gives the correction for the cell non-orthogonality.

Since the variables are required at both the cell centres and the cell face centres, the gradients of the variables at these locations are also needed. Gauss's theorem and the midpoint rule integral approximation is a simple and efficient way to obtain second order accuracy.

$$\int_V \mathrm{grad}\,\psi\, \mathrm{d}V = \int_S \psi\, \mathrm{d}\mathbf{s} \Rightarrow \quad (\mathrm{grad}\,\psi)_{P_0} \approx \frac{1}{V_{P_0}} \sum_{j=1}^{n_f} \psi_j\, \mathbf{s}_j \tag{C.7}$$

Here, ψ_j is the value of variable ψ at the cell face centre.

The first term in the prototype equation is different to the others because it contains an integral with respect to time. If the equation is rearranged into the following form:

$$\frac{d\Psi}{dt} = F(\phi) \tag{C.8}$$

where

$$\Psi = \int_V \rho B_\phi \mathrm{d}V \approx (\rho B_\phi V)_{P_0} \quad \text{and} \quad \phi = \phi(\mathbf{r},t)$$

then the left side of this equation is exactly integrated in the time interval between t_{m-1} and $t_m = t_{m-1} + \delta t_m$. However, the mean value of the right hand side, which incorporates the convective and diffusive fluxes and source terms, is approximated over the interval δt_m. This is done through a two-time-level implicit Euler scheme or a three-time-level implicit scheme. The first is a first-order fully implicit approximation that requires less computer memory then the second and imposes no limitations on the time step size. Its implicit form means that the current value of F is used for calculation of the quantity Ψ at t_m time:

$$\Psi^m = \Psi^{m-1} + F^m \delta t_m \tag{C.9}$$

The three-time-level implicit scheme extends the time integrating domain for one more time step and in the case of constant time steps it reads as:

$$\Psi^m = \Psi^{m-1} + \frac{2}{3} F^m \delta t_m + \frac{1}{3}\left(\Psi^{m-1} - \Psi^{m-2}\right). \tag{C.10}$$

It is accurate to the second order and is more stable than other schemes of the same accuracy. Switching to the Euler scheme, when accuracy is not essential, is straightforward.

Solution of the prototype equation for the moving boundaries requires implementation of the space conservation law. The mesh movement is known in advance for both the fluid flow and the solid body. The cell surface velocity $\mathbf{v}_s$ is calculated when either the two-time level or the three-time level scheme is used. It is more convenient to express that term as the sum of fluxes than to calculate the surface velocity separately. In the Euler implicit scheme the volume fluxes through the faces are:

$$\frac{V_{P_0}^m - V_{P_0}^{m-1}}{\delta t_m} = \sum_{j=1}^{n_f} \int_{S_j} \mathbf{v}_s \cdot \mathbf{ds} = \sum_{j=1}^{n_f} \frac{\delta V_j^m}{\delta t_m} = \sum_{j=1}^{n_f} \delta \dot{V}_j^m \tag{C.11}$$

while in the three-time-level implicit scheme, the face volume fluxes are:

$$\frac{V_{P_0}^m - 4V_{P_0}^{m-1} + V_{P_0}^{m-2}}{\delta t_m} = \sum_{j=1}^{n_f} \int_{S_j} \mathbf{v}_s \cdot \mathbf{ds} = \sum_{j=1}^{n_f} \frac{\delta V_j^m}{\delta t_m} + \sum_{j=1}^{n_f} \frac{\delta V_j^m - \delta V_j^{m-1}}{2\delta t_m} \tag{C.12}$$

The values $V_{P_0}^m, V_{P_0}^{m-1}, V_{P_0}^{m-2}$ are the volumes of the cell being calculated at times t_m, t_{m-1} and t_{m-2} respectively and δV_j^m and δV_j^{m-1} are the volumes swept by the cell face s_j during the two consecutive time intervals δt_m. This is an arrangement which allows the volume fluxes at the cell faces to be calculated without exactly knowing the value of the cell face velocity. The swept volumes are calculated when the coordinates of the vertices are known.

Transient Term

Transient rate of change is discretised through either the two-time-level Euler discretisation (C.9) or the three-time-level implicit discretisation scheme (C.10).

If the two-time-level implicit Euler scheme is used for approximation of the transient term, it becomes:

$$\frac{d}{dt} \int_V \rho B_\phi dV \approx \frac{(\rho B_\phi V)_{P_0} - (\rho B_\phi V)_{P_0}^{m-1}}{\delta t_m} \tag{C.13}$$

Assuming constant time steps, the transient term for the three-time-level implicit scheme becomes:

$$\frac{d}{dt} \int_V \rho B_\phi dV \approx \frac{3(\rho B_\phi V)_{P_0} - 4(\rho B_\phi V)_{P_0}^{m-1} + (\rho B_\phi V)_{P_0}^{m-2}}{\delta t_m} \tag{C.14}$$

The transient term has the same form for all dependent variables in the fluid flow model where B_ϕ stands for c_i, v_i, h, k, ε and only differs in the momentum equation for solids.

Convective Flux

The convective flux of variable ϕ through the cell face j in the prototype equation is a nonlinear term and it must first be linearized. This is done by the Picard iteration approach which assumes that the mass flux is known for the calculation of the variable, after which it is corrected by an iterative procedure until the iteration criterion is satisfied. The convective term then becomes:

$$C_j = \int_{S_j} \rho \phi (\mathbf{v} - \mathbf{v}_s) \cdot d\mathbf{s} \approx \dot{m}_j \phi_j^* . \tag{C.15}$$

The mass flux through the cell face $\dot{m}_j$ is defined as:

$$\dot{m}_j = \int_{S_j} \rho (\mathbf{v} - \mathbf{v}_s) \cdot d\mathbf{s} \approx \rho_j^* \left(\mathbf{v}_j^* \cdot \mathbf{s}_j - \dot{V}_j \right), \tag{C.16}$$

where the density ρ_j^* and velocity $\mathbf{v}_j^*$ are calculated through the pressure correction procedure described later in this appendix (C.29).

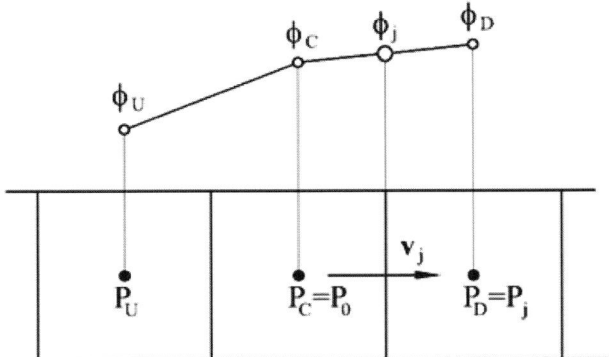

Figure A- 3 Upwind, central and downwind cell arrangement

There are more possibilities for the calculation of the value at the cell face centre ϕ_j^* in equation (C.15). To investigate them, variables at the centre of the cells and at the cell face centre are defined according to Figure A-3. A criterion to avoid non-physical oscillations is that the value $\phi_U \leq \phi_C \leq \phi_D$ or $\phi_D \leq \phi_C \leq \phi_U$ in the central cell must be locally bounded between upstream and downstream values. Its value depends on the flow direction.

The normalised variables at the end face α_j and at the centre of the cell α_C are calculated as

$$\alpha_j = \frac{\phi_j - \phi_U}{\phi_D - \phi_U}; \quad \alpha_C = \frac{\phi_C - \phi_U}{\phi_D - \phi_U} \tag{C.17}$$

The criterion for the appearance of non-physical oscillations α_j in the function of α_C is presented in Figure A-4. α_j for each numerical cell falls in the hatched region of the upper triangle if $0 \leq \alpha_C \leq 1$ and lies on the line $\alpha_j = \alpha_C$ if the numerical solution is bounded. Fulfilment of this criterion is especially important if the physical property may not be negative, like density or concentration, or when it may not exceed unity, as in the case of concentration. However, many factors affect these conditions like the presence of sources in the equations. Three schemes are mentioned here for the solution of the convective flux.

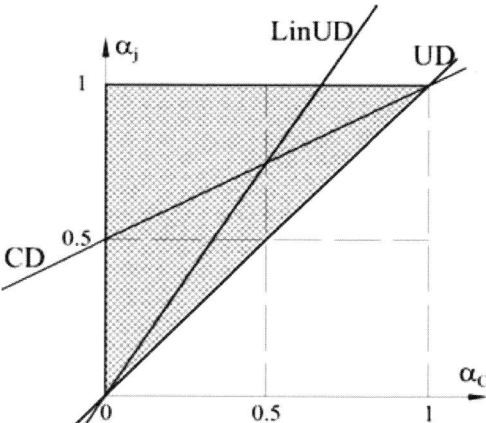

Figure A-4 Normalised variable diagram NVD

Upwind differencing is the scheme in which the value at the cell faces is equal to the value at the node upstream of the face. The value of the variable at the boundary ϕ_j^* is defined as:

$$\left(\phi_j^*\right)^{UD} = \begin{cases} \phi_{P_0} & \text{if} \quad (\mathbf{v} \cdot \mathbf{n}_j) > 0 \\ \phi_{P_j} & \text{if} \quad (\mathbf{v} \cdot \mathbf{n}_j) < 0 \end{cases} \tag{C.18}$$

Upwind differencing is the only approximation that satisfies the bounding criterion because $\alpha_j = \alpha_C$ and it never leads to an oscillatory solution. However, it imposes numerical diffusion on the calculation and results in first order accuracy. The values ϕ_{P_0} and ϕ_{Pj} are specified for the computational point and the centre of the neighbouring cell respectively.

Linear upwind differencing corrects the upwind value by the gradient of the variable. This accounts only for the upstream neighbours. The same rule as for the upwind differencing is applied. The cell face value is:

$$\left(\phi_j^*\right)^{LUD} = \begin{cases} \phi_{P_0} + \left(\mathrm{grad}\phi\right)_{P_0}^* \cdot \left(\mathbf{r}_j - \mathbf{r}_{P_0}\right) & \text{if} \quad (\mathbf{v} \cdot \mathbf{n}_j) > 0 \\ \phi_{P_j} + \left(\mathrm{grad}\phi\right)_{P_0}^* \cdot \left(\mathbf{r}_j - \mathbf{r}_{P_j}\right) & \text{if} \quad (\mathbf{v} \cdot \mathbf{n}_j) < 0 \end{cases} \quad (C.19)$$

The scheme has a second order value of accuracy because of the linear interpolation between points but it is only conditionally bounded because $\alpha_j = 1.5\alpha_C$ as presented Figure A-4.

The central differencing scheme is similar to the linear upwind but the value at the cell face is obtained from linear interpolation between the values at the calculating and neighbouring cells. The value of the variable at the cell face has the form of equation (C.6) and is written as:

$$\phi_j^{CD} = \frac{1}{2}\left(\phi_{P_0} + \phi_{P_j}\right) + \frac{1}{2}\left[\left(\mathrm{grad}\phi\right)_{P_0} \cdot \left(\mathbf{r}_j - \mathbf{r}_{P_0}\right) + \left(\mathrm{grad}\phi\right)_{P_j} \cdot \left(\mathbf{r}_j - \mathbf{r}_{P_j}\right)\right] \quad (C.20)$$

Gradients in this equation are calculated from the values in both the central and neighbouring cells. The central differencing scheme is of second order of accuracy and, similar to the linear upwind, it can give unbounded and oscillatory solutions. This happens if the computational grid is too coarse. However, once the grid is sufficiently fine, the result converges faster than in other schemes.

There are other schemes which combine these three basic ones and take advantage of each of them. For example, the *MINMOD* scheme is always of second order of accuracy and it uses upwind differencing in the unbounded regions when $\alpha_C < 0$ and $\alpha_C > 1$, linear upwind in the region $0 < \alpha_C < 0.5$ and central differencing for $0.5 < \alpha_C < 1$. There are also some recent schemes, which are incorporated in commercial software, like that of *Przulj and Basara* (2001).

To prevent non-physical oscillations and wiggles, a combination i.e. a blending of the second and first order schemes is often used. It is based on the following formula:

$$\phi_j^* = \phi_j^{FO} + \gamma_\phi \left(\phi_j^{SO} - \phi_j^{FO}\right) \quad (C.21)$$

where, ϕ_j^{FO} is the value obtained by the scheme of first order accuracy, while ϕ_j^{SO} represents the second order scheme. The blending factor γ_ϕ is a constant for the calculating domain but its value depends on the mesh quality.

Diffusive Flux

The diffusive flux of the variable ϕ through the internal cell face j is approximated by the use of the mid point rule approximation of the surface integral.

$$D_j = \int_{S_j} \Gamma_\phi \left(\mathrm{grad}\,\phi\right)^*_{P_0} \cdot \mathrm{d}\mathbf{s} \approx \Gamma_{\phi j} \left(\mathrm{grad}\,\phi\right)^*_j \cdot \mathbf{s}_j \qquad (C.22)$$

In this equation $\Gamma_{\phi j}$ stands for diffusivity at the cell face centre, obtained by using (C.6). However, a second order approximation of the gradient from (C.7) cannot handle oscillations which have a period of twice the characteristic length of the numerical mesh and due to that, a third order dissipative term should be added to the interpolated value.

Source Terms

The source terms on the right hand side of the prototype equation consist of the surface and volume integrals. They are discretised by means of the midpoint rule (C.4). As a result of discretisation, the surface integral over all the faces of the numerical cell becomes:

$$Q_{\phi S} = \int_S \mathbf{q}_{\phi S} \cdot \mathrm{d}\mathbf{s} \approx \sum_{j=1}^{n_f} \mathbf{q}_{\phi S} \cdot \mathbf{s}_j \qquad (C.23)$$

while the volume integral in the CV becomes:

$$Q_{\phi V} = \int_V q_{\phi V} \mathrm{d}V \approx \left(q_{\phi V}\right)_{P_0} V_{P_0} \qquad (C.24)$$

After the necessary values for the prototype equation are set, the boundary and initial conditions are implemented and a procedure for pressure-velocity coupling has to be applied before the system of algebraic equations can be solved.

Boundary and Initial Conditions

Boundary conditions on the cell faces coinciding with the boundary of the solution domain are applied prior to solution of the system of algebraic equations. All boundaries to the fluid flow through which it is connected to the solid parts are no-slip walls with either, known temperature or a temperature approximated by the earlier explained procedure. Due to that, a cell face flux ϕ_j^* becomes a boundary flux ϕ_B in all equations where boundary cell faces are considered. In such a case, the mass flux in the momentum equation at the boundary is zero, the heat flux through the boundary for the energy equation is calculated from the wall temperature and the thermal conductivity in the near wall region (B.7), while the concentration flux reads zero. Diffusive fluxes are also replaced with their boundary values.

Since the compressor flow is always in a transient state, unsteady calculation is necessary. This, in turn, requires the initial conditions to be prescribed for the dependent variables at each control volume of the computational domain. A proper estimation of these plays a significant role in the efficiency and computational time required to obtain consistent results. The calculation is complete if the time history for all dependant variables is equal for two consecutive cycles.

The initial values of the velocities in the momentum equation are set to zero in all the cells within the working chamber. The initial pressure is prescribed for the cells at the inlet and outlet receiver as the inlet and outlet pressure. For all other cells the initial values are calculated as a linear interpolation between these values with respect to the relative distance in the z direction as:

$$p_i^0 = p_{inl}^0 + \frac{z_i}{L}\left(p_{out}^0 - p_{inl}^0\right). \tag{C.25}$$

z_i is the cell centre distance starting from the coordinate origin, while L is the overall compressor length. This simple method to prescribe the initial values often gives a consistent final solution within 4 to 5 compressor cycles. The initial temperature is calculated in the same manner as the pressure, i.e. as the linear interpolation between the prescribed inlet and outlet temperatures T^0_{inl} and T^0_{out}. The density is then calculated from the equation of state. The concentration is also interpolated between the prescribed values at the inlet and the outlet of the compressor in a similar manner to the other variables. Initial values of the kinetic energy and its dissipation rate are set at zero throughout the domain.

When the Euler implicit time integration is employed, these prescribed values at time t_0 are sufficient for the calculation. If, however, the three time level implicit scheme is used, the values at the time $t_{-1}=t_0-\delta t$ should be given. They are set at the same as values at time t_0.

Derived System of Algebraic Equations

Discretisation of the derived algebraic equation results in the same form for all variables:

$$a_{\phi 0}\phi_{P_0} - \sum_{j=1}^{n_j} a_{\phi j}\phi_{P_j} = b_\phi, \tag{C.26}$$

where the index 0 determines the control volume in which the variable is calculated and the index j defines the neighbouring cells. The number n_i represents the internal cell faces between the calculating cell and the neighbouring cells. The right hand side terms are known from either the previous iteration or the time step. All the coefficients, central $a_{\phi 0}$, neighbouring $a_{\phi j}$ and right hand side b_ϕ, are treated explicitly to increase computational efficiency.

$$a_{\phi j} = \Gamma_{\phi j} \frac{\mathbf{s}_j \cdot \mathbf{s}_j}{\mathbf{d}_j \cdot \mathbf{s}_j} - \min(\dot{m}_j, 0),$$

$$a_{\phi 0} = \sum_{j=1}^{n_f} a_{\phi j} + \frac{(\rho V)_{P_0}^{m-1}}{\delta t_m},$$

$$b_\phi = \sum_{j=1}^{n_f} \Gamma_{\phi j} \left((\operatorname{grad} \phi)_j \cdot \mathbf{s}_j - \overline{\operatorname{grad} \phi} \cdot \mathbf{d}_j \frac{\mathbf{s}_j \cdot \mathbf{s}_j}{\mathbf{d}_j \cdot \mathbf{s}_j} \right) -$$

$$\sum_{j=1}^{n_f} \frac{\gamma_\phi}{2} \dot{m}_j \left((\mathbf{r}_j - \mathbf{r}_{P_0}) \cdot (\operatorname{grad} \phi)_{P_0} + (\mathbf{r}_j - \mathbf{r}_{P_0}) \cdot (\operatorname{grad} \phi)_{P_j} + (\phi_{P_j} - \phi_{P_0}) \frac{\dot{m}_j}{\operatorname{abs}(\dot{m}_j)} \right) +$$

$$Q_{\phi S} + Q_{\phi V} + \sum_{B=1}^{n_B} a_{\phi B} \phi_B + \frac{(\rho V \phi)_{P_0}^{m-1}}{\delta t_m}.$$

(C.27)

$\mathbf{d}_j$ is a distance vector, which is effective if the mesh is non-orthogonal. It is then used to correct the cell face value. It is the normal distance between the line connecting two neighbouring cell points and the cell face centre. n_B is the number of boundary faces surrounding the cell P_0. The coefficient $a_{\phi B}$ for the centre point at the boundary cell face is calculated in a similar manner to that of the neighbouring coefficient $a_{\phi j}$, assuming the distance between the boundary point and the centre of the cell.

Pressure Calculation

Pressure has no governing equation and a special procedure is developed for its calculation. It is performed in three steps. Firstly the velocity and density fields are obtained from the momentum equation regardless of whether the continuity equation is satisfied. Then a pressure correction is calculated to satisfy the continuity equation in the predictor step and finally a corrector stage is applied in which new values of the velocity, pressure and density fields are calculated. This procedure is called SIMPLE algorithm. The velocity through the cell face is calculated to take into account the pressure diffusion:

$$\mathbf{v}_j^* = \mathbf{v}_j - \left(\frac{V_{P_0}}{a_{v_0}}\right)\left\{\frac{p_{P_j}-p_{P_0}}{|\mathbf{d}_j|} - \frac{\overline{\mathrm{grad}p}\cdot\mathbf{d}_j}{|\mathbf{d}_j|}\right\}\frac{|\mathbf{d}_j|\,\mathbf{s}_j}{\mathbf{d}_j\cdot\mathbf{s}_j} \qquad (C.28)$$

The first term in this equation is the cell face velocity obtained by the use of (C.6), while the remainder is a third order pressure diffusion term. It acts as a correction to the interpolated velocity if oscillations in the pressure field are present. Otherwise it is negligible. This term vanishes if the pressure varies linearly or quadratically and in other cases it is proportional to the square of the mesh spacing. Therefore, it is reduced by grid refinement together with other discretisation errors of the second order. The value a_{v_0} in this term is the corresponding central coefficient of the momentum equation.

To enhance stability, the value of the density at the centre of the cell face is calculated as a blend of the first and second order interpolations in the same manner as in (C.21). The value of the blending factor is usually high, $\gamma_r \approx 0.95$:

$$\rho_j^* = \rho_j^{UD} + \gamma_\rho\left(\rho_j^{CD} - \rho_j^{UD}\right) \qquad (C.29)$$

The pressure correction equation is now constructed to satisfy the momentum equation in which predictor stage values $\mathbf{v}^{\text{pred}}$, p^{pred} and ρ^{pred} featuring in (C.28) and (C.29) also satisfy the continuity equation. If the Euler implicit time differencing scheme is employed the continuity equation can be written in a form convenient for further calculation as:

$$\frac{(\rho V)_{P_0} - (\rho V)_{P_0}^{m-1}}{\delta t_m} + \sum_{j=1}^{n_f} \dot{m}_j = 0 \qquad (C.30)$$

The pressure correction equation is now:

$$a_{p_0}p'_{P_0} - \sum_{j=1}^{n_j} a_{p_j}p'_{P_j} = b_{p'}, \qquad (C.31)$$

with coefficients:

$$a_{p_j} = \rho_j^* \overline{\left(\frac{V_{P_0}}{a_{v_0}}\right)}\frac{\mathbf{s}_j\cdot\mathbf{s}_j}{\mathbf{d}_j\cdot\mathbf{s}_j} - \left[(1-\gamma_\rho)\min(\mathbf{v}_j^*\cdot\mathbf{s}_j,0)+\frac{1}{2}\gamma_\rho\mathbf{v}_j^*\cdot\mathbf{s}_j\right]\left(\frac{\partial\rho}{\partial p}\right)_{P_j}\beta_p,$$

$$a_{p_0} = \sum_{j=1}^{n_f}\hat{a}_{p_j} + \frac{V_{P_0}}{\delta t_m}\left(\frac{\partial\rho}{\partial p}\right)_{P_0},$$

$$b_{p'} = -\sum_{j=1}^{n_f}\dot{m}_j - \left((\rho V)_{P_0} - (\rho V)_{P_0}^{m-1}\right).$$

$$(C.32)$$

The central coefficient in (C.32) is obtained from 'conjugate' values $\hat{a}_{p_j}$ of the central coefficients at the neighbouring cells a_{p_j} if P_0 and P_j exchange their roles. The compressibility coefficient defined as $C_\rho = \partial\rho/\partial p$ is calculated in Section 2.2.4, while β_p is the under-relaxation factor for the pressure correction equation.

Finally, in the corrector stage, the velocity, pressure and density fields are corrected for the calculated value of pressure correction:

$$\mathbf{v}_{P_0} = \mathbf{v}_{P_0}^{pred} - \frac{1}{a_{v_0}} \sum_{j=1}^{n_f} p'_j \mathbf{s}_j,$$

$$p_{P_0} = p_{P_0}^{pred} + \beta_p p'_{P_0}, \quad (C.33)$$

$$\rho_{P_0} = \rho_{P_0}^{pred} + \left(\frac{\partial \rho}{\partial p}\right)_{P_0} \beta_p p'_{P_0}.$$

The mass fluxes which satisfy the continuity equation are calculated from the equation:

$$\dot{m}_j = \dot{m}_j^{pred} - a_{p_j} p'_{p_j} + \hat{a}_{p_j} p'_{p_0} \quad (C.34)$$

and these are used for computation of the convective fluxes in the next iteration.

Since the boundary conditions of the momentum equation are prescribed velocities, in the case of screw compressor flow, and these are Dirichlet boundary conditions, then the zero gradient boundary conditions on the pressure correction are applied.

The pressure correction equation adjusts itself automatically to the type of the flow. In the region of low Mach numbers the contribution of the density correction is small. In regions of Mach number close and higher then 1, the contribution of the density correction becomes dominant and the equation becomes hyperbolic contrary to the previous case, when it was elliptic. This feature is very important for screw compressor flows in which both low and high Mach number regions are encountered.

References

Bradshaw P, 1994: Turbulence: The Chief Outstanding Difficulty of Our Subject, Experiments in fluids, Vol 16, 203-216
Chawner J. R, Anderson D. A, 1991: Development of an Algebraic Grid Generation Method with Orthogonality and Clustering Control, Conference on Numerical Grid Generation in CFD and Related Fields at Barcelona, 107
Demirdzic I, Peric M, 1990: Finite Volume Method for Prediction of Fluid flow in Arbitrary Shaped Domains with Moving Boundaries, Int. J. Numerical Methods in Fluids Vol.10, 771
Demirdzic I, Lilek Z, Peric M, 1993: A Collocated Finite Volume Method for Predicting Flows at All Speeds, Int. J Numerical Methods in Fluids, Vol. 16, 1029
Demirdzic I, Muzaferija S, 1995: Numerical Method for Coupled Fluid Flow, Heat Transfer and Stress Analysis Using Unstructured Moving Mesh with Cells of Arbitrary Topology, Comp. Methods Appl. Mech Eng, Vol.125 235-255
Eiseman P. R, 1985: Grid Generation for Fluid Mechanics Computations, Annual Rev. Fluid Mech, Vol.17, 487-522
Eiseman P. R, 1991: Control Point Forms for Interactive Grid Manipulation, Computer Methods in Applied Mechanics and Engineering, Vol.91, 1151-1156
Eiseman P. R, 1992: Control Point Grid Generation, Computers Mathematical Applications, Vol.24, No.5/6, 57-67
Eiseman P. R, Hauser J, Thompson J.F, Waterhill N.P, 1994: (Ed.) Numerical Grid Generation in Computational Field Simulation and Related Fields, Proceedings of the 4th International Conference, Pineridge Press, Swansea Wales, UK
Ferziger J. H, Peric M, 1996: Computational Methods for Fluid Dynamics, Springer, Berlin
Field D. A, 2000: Qualitative Measures for Initial Meshes, International Journal for Numerical Methods in Engineering, Vol.47, 887-906
Fletcher C. A. J, 1991: Computational techniques for fluid dynamics, vol. I, Springer, Berlin
Gordon W. J, 1969: Distributive lattices and the approximation of multivariate functions. Symposium on Approximation with Special Emphasis on Spline Functions, Academic Press, Madison, 223-277
Gordon W. J, Hall C. A, 1973: Construction of Curvilinear Coordinate Systems and Aplications to Mesh Generation, Int. J. Numer. Meth. Engineering, Vol.7, 461-477
Gordon W. J, Thiel L. C, 1982: Transfinite Mappings and Their Application to Grid Generation. In Thompson J.F (ed.): Numerical Grid generation, North Holland, 171-192
Hanjalic K, 1994: Advanced Turbulence Closure Models: A View on the Current Status and Future Prospects, Int. J. Heat & Fluid Flow, Vol 15, 178-203
Kim J. H, Thompson J. F, 1990: 3-Dimensional Adaptive Grid generation on a Composite-Block Grid, AIAA Journal, Vol.28, Part.3, 470-477
Knupp P, Steinberg S, 1993: Fundamentals of Grid Generation, CRC Press, Boca Raton

Kovacevic A, Stosic N, Smith I. K, 1999: Development of CAD-CFD Interface for Screw Compressor Design, International Conference on Compressors and Their Systems, London, IMechE Proceedings, 757

Kovacevic A, Stosic N, Smith I. K, 2000: The CFD Analysis of a Screw Compressor Suction, International Compressor Engineering Conference at Purdue, 909

Kovacevic A, Stosic N, Smith I. K, 2000: Grid Aspects of Screw Compressor Flow Calculations, Proceedings of the ASME Advanced Energy Systems Division – 2000, Vol. 40, 83

Kovacevic A, Stosic N, Smith I. K, 2001: CFD Analysis of a Screw Compressor Performance, International Conference on Compressors and Their Systems, London, IMechE Proceedings

Kovacevic A, Stosic N, Smith I. K, 2001: Analysis of Screw Compressor Performance by means of 3-Dimensional Numerical Modelling, Seminar on Advances of CFD in Fluid Machinery Design, London, IMechE Proceedings

Kovacevic A, Stosic N, Smith I. K, 2003: Three Dimensional Numerical Analysis of Screw Compressor Performance, Journal of Computational Methods in Science and Engineering, vol. 3, no. 2, pp. 259- 284

Kovacevic A, Stosic N, Smith I. K, 2004: A Numerical Study of Fluid-Solid Interaction in Screw Compressors, International Journal of Computer Applications in Technology (IJCAT), Vol. 21, No. 4, 2004, pp. 148-158

Kovacevic A 2005: Boundary Adaptation in Grid Generation for CFD Analysis of Screw Compressors, International Journal for Numerical Methods in Engineering (IJNME), vol. 63, 2005

Lehtimaki R, 2000: An Algebraic Boundary Orthogonalisation Procedure for Structured Grids, International Journal for Numerical Methods in Fluids, Vol. 32, 605-618

Lin W. L, Chen C. J, 1998: Automatic Grid Generation of Complex Geometries in Cartesian Co-ordinates, International Journal for Numerical methods in Fluids, Vol.28, No9, 1303-1324

Liou Y. C, Jeng Y. N, 1995: Algebraic Method Using Grid Combination, Numerical heat Transfer, Part B, Vol.28, 257-276

Liseikin V. D, 1991: Techniques for Generating Three-dimensional Grids in Aerodynamics (review). Problems Atomic Sci. Technology. Ser. Math. Model. Phys Process Vol.3, 31-45 (in Russian)

Liseikin V. D, 1996: Construction of Structured Adaptive Grids – a Review, Comp. Math. Math Phys, Vol.36, No.1, 1-32

Liseikin V. D, 1998: Algebraic Adaptation Based on Stretching Functions, Russian Journal for Numerical and Analytical Mathematical Modelling, Vol.13, No.4, 307-324

Liseikin V. D, 1999: Grid generation Methods, Springer-Verlag

Lumley L. J, 1999: Engines - an introduction, Cambridge University press, UK

Moitra A, 1992: Two- and Three-Dimensional Grid Generation by an Algebraic Homotopy Procedure, AIAA Journal Vol.30, No.5, 1433

Oden J. T, 1972: Finite elements of non-linear continua, McGraw Hill, New York

Owen S. J, 1998: A Survey of Unstructured Mesh Generation Technology, published on web http://www.andrew.cmu.edu/user/sowen/survey/index.html

Owen S. J, Staten M. L, Cannan S. A, Saigal S, 1998: Quad-Morphing: Advanced Front Quad Meshing Using Triangle Transformations, http://www.andrew.cmu.edu/user/sowen/qm/rt98quad.htm

Patankar S. V, 1980: Numerical Heat Transfer and Fluid Flow, McGraw Hill, London

Peric M, 1990: Analysis of pressure-velocity coupling on non-orthogonal grids, Numerical Heat Transfer, Part B, 17, 63-82

Rinder L, 1984: Schraubenverdichterlaeufer mit Evolventenflanken (Screw Compressor Rotor with Involute Lobes), Proc. VDI Tagung "Schraubenmaschinen 84" VDI Berichte Nr. 521 Duesseldorf

Saha S, Basu B. C, 1991: Simple Algebraic Technique for Nearly Orthogonal Grid Generation, AIAA Journal, Vol.29, No.8, 1340

Samareh-Abolhassani J, Smith R. E, 1992: A Practical Approach to Algebraic Grid Adaption, Computers Mathematical Applications, Vol.24, No.5/6, 69-81

Shih T. I. P, Bailey R. T, Ngoyen H. L, Roelke R. J, 1991: Algebraic Grid Generation For Complex Geometries, International Journal for Numerical Methods in Fluids, Vol. 13, 1-31

Smith G. D, 1985: Numerical Solution of Partial Differential Equations: Finite Difference Methods, 3rd edition, Claredon press, Oxford

Smith R. E, 1982: Algebraic Grid Generation, from Numerical Grid Generation, Ed. By Thomson J.E, Elsevier Publishing Co, 137

Smith R. E, 1992: (Ed.) Proceedings of the Software Systems for Surface Modelling and Grid Generation Workshop, NASA Conference Publication 3143, Nasa Langley Research Centre, Hampton, VA

Smith R. E, Johnson L. J, 1996: Automatic Grid Generation and Flow Solution for Complex Geometries, AIAA Journal, Vol.34, No.6, 1120-1124

Soni B. K, 1992: Grid Generation for Internal Flow Configurations, Computers Mathematical Applications, Vol.24, No.5/6, 191-201

Soni B.K, Thompson J. F, Eiseman P. R, Hauser J, 1996: (Eds.) Numerical Grid Generation in Computational Filed Simulation, Proceedings of the 5th International Conference, MSU Publisher, Mississippi State, MS, USA

Steinthorsson E, Shih T. I. P, Roelke R. J, 1992: Enhancing Control of Grid Distribution In Algebraic Grid generation Generation, International Journal for Numerical Methods in Fluids, Vol. 15, 297-311

Stosic N., Milutinovic Lj., Hanjalic K. and Kovacevic A, 1992: Investigation of the Influence of Oil Injection upon the Screw Compressor Working Process, Int.J.Refrig. 15,4,206

Stosic N, Smith I. K. and Zagorac S, 1996: CFD Studies of Flow in Screw and Scroll Compressors, XIII Int. Conf on Compressor Engineering at Purdue

Stosic N, 1998: On Gearing of Helical Screw Compressor Rotors, Proc IMechE, Journal of Mechanical Engineering Science, Vol.212, 587

Stosic N, Smith I. K, Kovacevic A, 2003: Opportunities for Innovation with Screw Compressors, Proceedings of IMechE, Journal of Process Mechanical Engineering, Vol 217, pp 157-170, 2003

Stosic N, 2004: Screw Compresors in Refrigeration and Air Conditioning, Int Journal of HVACR Research, 10(3) pp 233-263, July 2004

Stosic N, Smith I. K, Kovacevic A, 2005: Screw Compressors Mathematical Modelling and Performance Calculation, ISBN-10 3-540-24275-9, Springer Berlin Heidelberg New York

Thomas D. P and Lombard K. C, 1979: Geometric Conservation Law and its Application to Flow Computations on Moving Grids, AIAA Journal, 17, 1030-1037

Thompson J. E, 1984: Grid Generation Techniques in Computational fluid Dynamics, AIAA Journal, Vol.22, No.11, 1505-1523

Thompson J. E, Warsi Z. U. A, Martin C. W, 1985: Numerical Grid Generation – Foundations and Applications, Elsevier Science Publishing Co, also J.F. Thompson web edition 1997 http://www.erc.msstate.edu/education/gridbook

Thompson J. E, Waterhill N. P, 1993: Aspects of Numerical Grid Generation: Current Science and Art, AIAA Journal, Vol.93, 3539

Thompson J. E, 1996: A Reflection on Grid Generation in the 90's: Trends, Needs and Influence. In Soni B.K, Thompson J.F, Hauser J, Eiseman P.R (eds.): Numerical grid Generation in CFD, Mississippi State University, Vol.1, 1029-1110. Also published on web 2000, http://www.erc.msstate.edu/~joe/gridconf

Thompson J. E, Soni B, Weatherrill N. P, 1999: Handbook of Grid generation, CRC Press

Trulio G. J, Trigger R. K, 1961: Numerical solution of the one-dimensional hydrodynamic equations in an arbitrary time-dependent coordinate system, University of California Lawrence Radiation Laboratory Report UCLR-6522

Versteeg K. H, Malalasekera W, 1995: An Introduction to computational Fluid Dynamics – The Finite Volume Method, Longman Scientific and Technology, UK

Vinokur M, 1983: On One-Dimensional Stretching Functions for Finite-Difference Calculations, Journal of Computational Physics, Vol.50, 215-234

Zhu J, Rodi W, Schoenung B, A Fast Method For Generating Smooth Grids, 1998

Printing: Krips bv, Meppel
Binding: Stürtz, Würzburg